Dear

MATH

WHY KIDS HATE MATH AND WHAT TEACHERS CAN DO ABOUT IT

SARAH STRONG & GIGI BUTTERFIELD

TIMES
TEN

Dear Math
© 2022 by Times 10 Publications
Highland Heights, OH 44143 USA
Website: 10publications.com

All web links in this book are correct as of the publication date but may have become inactive or otherwise modified since that time.

Cover and Interior Design by Steven Plummer
Editing by Jennifer Zelinger Marshall
Copyediting by Jennifer Jas

Paperback ISBN: 978-1-956512-18-2
eBook ISBN: 978-1-956512-19-9
Hardcover ISBN: 978-1-956512-20-5

Library of Congress Cataloging-in-Publication Data is available for this title.

First Printing: June 2022

PRAISE FOR *DEAR MATH*

"I deeply appreciate the mixture of perspectives from students and their teacher as the author integrates her own experience as a math teacher with the thinking of prominent math education scholars and the energy of math Twitter. The book feels like a familiar conversation about math reform in our country, suffused with the enthusiasm (and pain) of the voices of young people. It turns out the kids have some thoughts on their math education."

— BEN DALEY, PRESIDENT, HIGH TECH HIGH
GRADUATE SCHOOL OF EDUCATION

"In centering the voices of high school students, *Dear Math* practices what it preaches: a radically empathic approach to a school subject that too often leaves damage and destruction in its wake. Strong and Butterfield offer a narrative that is conceptually rich and also deeply useful, identifying a range of mindsets and practices that can help transform mathematics classrooms into spaces replete with curiosity, collaboration, and deep learning. Teachers and school leaders alike will welcome the book's rare combination of readability, usability, and depth."

— SARAH FINE, IN SEARCH OF DEEPER LEARNING

"A tremendous amount of time and energy go into attempts to improve the education system. What is unfortunate is that rarely are the students given an opportunity to contribute their thoughts and ideas. Sarah Strong and Gigi Butterfield have made an important and valuable contribution to the field: a book dedicated to giving students a voice and letting them tell their story about mathematics. The results are powerful. This should be required reading for every math teacher."

— DR. PATRICK CALLAHAN, CALLAHAN CONSULTING

"Strong and Butterfield's book centers student voice and the power of teaching in ways you have never seen before. The Dear Math letters they share will take you inside the minds of students who dislike math and (wrongly!) believe that math dislikes them, all before offering you practical strategies for restoring and rehabilitating that relationship."

— DAN MEYER, DESMOS

This book is dedicated to all the middle and high school students with whom we have shared semesters and school years. Thank you for your ideas, your light, and your brilliance.

TABLE OF CONTENTS

FOREWORD

DEAR **R**EADER,

I am writing to you in early spring 2022. It has been an especially difficult couple of years for students and for teachers. School closures, distance learning, school re-openings, and protocols require us to be distanced in order to be together. These events—on top of the already exhausting work of teaching—have put additional strain on individual teachers, on students, on schools, and on school systems.

Where do we find hope for a better future in schools? Answers will vary, of course. But for many of us, this hope lives in stories.

Sarah and Gigi have written a book chock full of stories, and these stories are hopeful because they center on connections. They tell of students connecting math to the places where they find meaning and beauty, such as Yosef in Chapter 7, who sees geometric transformations as birds flying freely in the air. In Chapter 6, they introduce us to the work of a colleague, Chris Nho, who connects students' mathematical work to the ways mathematics is used outside of classrooms with a simple phrase recognizing accomplishment: "Wow, that's a real professional mathematician move right there!" Sarah and Gigi model for us a lovely connection between teacher and student, in which they had the opportunity to build worlds together.

At its heart, *Dear Math* is a book about listening to students, and it reminds me of the inspiring tradition of centering students' ideas, including the research work of Cognitively Guided Instruction (CGI) and of Integrating Mathematics and Pedagogy (IMAP), as well as the instructional and professional work of the Math Forum. While at the Math Forum, Max Ray-Riek talked about two being greater than four because listening *to* students is more powerful than listening *for* particular answers. *Dear Math* letters stand proudly among the practices studied and described in these projects.

Throughout *Dear Math*, Sarah and Gigi connect us to contemporary mathematicians and math teachers writing and working in a wide range of media, from books to TikTok. These folks share a common goal of listening in order to better understand and to better connect—and in doing so, to make schools better places for learning mathematics.

As you read this book, I trust that you will be inspired by the bravery, honesty, and originality of the students who wrote the letters you'll read. Additionally, you will feel connected to a wide and wise network of mathematics teachers and learners. I am certain you will feel refreshed and hopeful.

Sincerely yours,

Christopher Danielson, PhD
Curriculum Developer at Desmos; Founder of Math On-A-Stick
talkingmathwithkids.com

INTRODUCTION

By Sarah Strong

I **FIRST LEARNED THE** term "worldbuilding" on a podcast in the summer of 2021. The podcast was not about math or math education, but I immediately adopted the term into my way of thinking about my work. Worldbuilding is a science fiction and fantasy writing concept where the author designs an imaginary world within a fictional, or sometimes real, world. I would like to conceive of this book as a worldbuilding book. Before you put this book down because you think it might be from a genre you aren't interested in, let me explain.

Some early feedback I received about this book was that it wasn't relatable and that my context at a small project-based learning school in Southern California situated me as too much of an outlier in the world of mathematics education. I took that feedback seriously. As I wrote, I continually tried to imagine how folks from seemingly "opposite" school systems might think of this work. I considered the ways my school was different from other schools. In my school, teachers are liberated to teach as they see fit. For better or worse, teachers can try and fail and then try again the next year. The school is a charter school and has used a ZIP code-based lottery system from around the diverse county since its inception. It has never had academic tracking at any grade level, there are no AP or IB courses, the students are on interdisciplinary teaching teams, teachers are designers of larger projects and daily lessons, and teachers are given autonomy, freedom, and support to try lots of different approaches. The students in this space, while necessarily representing the demographics of San Diego as a whole, are given an incredible voice to share what does and doesn't work for them through the consistent use of shared schoolwide data collection measures, exit tickets, and advisory groups, and by playing integral parts in the hiring process of teachers. They create project work that is purposefully beautiful and often for an audience outside of the school building. More importantly, our classrooms are places of discourse, where we celebrate diversity, wrestle with tough issues, and consider it necessary to work with new groups of students.

Another piece of the puzzle leading to the co-authorship of this book was that I happened to teach Gigi for four years in a row. I didn't set out to do this. I had been a ninth grade teacher for five years and absolutely loved it. During Gigi's sophomore year, however, a tenth grade teacher moved on, and I was asked to step into her classes, which included Gigi. After that year, an eleventh and twelfth grade combo class opened up, and I was asked to take on this challenging situation; Gigi was one of the eleventh graders in this cohort. Finally, the pandemic lockdown happened during Gigi's senior year, and I continued to teach twelfth graders online to support them in the transition of their final year. By pure happenstance, Gigi was once again in my class.

While this is uncommon, even in our school, it afforded us a lot of time to explore math and contemplate education together. We were worldbuilding in every conversation! We didn't set out to write a book about it, but at the beginning of Gigi's junior year, another teacher walked in on one of our deep conversations about math teaching and learning and said, "This is fascinating; you should write a book!"

We looked at each other and laughed, and then we continued to joke about it for another six months until one day, we decided the idea might actually be worth pursuing.

Whether you work in a tiny religious school in rural Idaho; a large, comprehensive high school in Memphis; or PS138 in New York City, I believe with all my heart that many of the pieces that make my school unique can and must work in other places, not because of what we do but because of how we do it.

I believe unequivocally in the need for an education system grounded in equity, justice, and ending poverty. And that our class-room spaces, particularly our math classes, are places where teachers get to catch glimpses of this possibility every day. I believe that most of our spaces are not yet designed with such values. How might we carry the dreams of what could be and hold them in conjunction with the current reality? Here enters the idea of worldbuilding.

The world that I built in my unique math classroom involved a process of listening to students and designing with their voices and identifying markers in mind first—and connecting to broader research-backed practice and data second. I developed the fol-lowing new framework for how I might listen to students.

THE FOUR DIMENSIONS OF LISTENING TO STUDENTS:

1. Listen to students' mathematical thinking and ground daily learning goals in what they share.

2. Listen for the practices and habits of a mathematician and celebrate those regularly.

3. Listen for how students are doing or feeling in their lives outside of school each day.

4. Listen to students' stories in Dear Math letters in hopes of gaining a greater understanding of their mathematical identities.

In this book, we will define mathematical identity as the way a person sees themself as a mathematician and participates in mathematical activities with others. Continuing to posture ourselves, as teachers, with an eye toward learning, growth, and liberation will result in the best possible system for all schools. While I believed this in theory, it wasn't until I started diving into broader research in the field of mathematics and engaging in teacher research as a graduate student, and then as a member of teacher-researcher networks, that I came to truly believe this method can and *must* be relevant in all schools.

I developed the practices I share in this book in conjunction with large groups of students and teachers in my communities. I received kind, specific, and helpful feedback that guided me to adapt, adopt, or abandon whatever idea I was trying on a given day. I ran dozens of research cycles and hundreds of "looking at student work" protocols and exit ticket data collection, and I arrived at the practices in this book that continue evolving to this day. As frequently as possible, I share the evolving work of these practices and invite you into their evolution.

Through all of this, however, I do not say I am a "researcher." I am a teacher in my classroom just trying to get better at getting better. I am trying to see how the research I read from my heroes, many of whom I mention throughout this book, looked in practice in my context and how my students experienced it. And I was fortunate enough to be at a school that let me try many approaches and bring my students into the process. Listening to students, however, is no easy task.

What do Dear Math letters have to do with all of this? The students who walk into our classrooms are wonderfully, beautifully complex beings. Even if we remove all the other pieces of their stories and just look at their math experiences, we will find a huge variety if we pull back the curtain and listen to their stories. For a long time,

I used the ignorance-is-bliss mentality when it came to my students' math stories. Yes, there may have been dark and troubled pasts, but I was there to save the day and teach them better than they had ever been taught before! Through masterful explaining, engaging lessons and projects, and a generally jolly attitude, I would be everything every child needed. I tried valiantly at this for a long time.

And it was exhausting.

Each child needs to achieve "something different" from a unique starting place—their skills, knowledge, and identity—by working at a unique pace.

A math problem started brewing in my head: "If thirty students walk into a room with thirty different math pasts, thirty different current mathematical needs, thirty different ways of thinking, and thirty different learning paces, how many different lessons do I need to plan on a given day?"

The result of that query was mind-numbing. Did I go to school for seven years just to take part in the most impossible job ever?

A few years into my teaching career, I started doing various types of math autobiographies and math letters to kick off the year. Students would pour out their hearts in these letters (or do the bare minimum to get by), lamenting past traumas, and (hopefully) forging a new path forward. I became intrigued and noted how many of them were linked to past math teachers and classes. Of course, we know the value of a teacher is extremely high, but I felt like I was throwing all of their past teachers under proverbial buses. I was sitting around with one of my humanities colleagues one day while she talked about having her students write "Dear Books" letters, and it dawned on me.

If I had my students write Dear Math letters, I could, as best as possible, come to understand their relationship with the discipline of math itself. I could hear how they speak to math, how they share formative parts of their math upbringing, how they speak about their future with the subject, and much more.

The process of picturing math as someone sitting next to you at the table that you can have a conversation with was exciting. And for students, the personification of the subject seemed to add personal appeal. As the students wrote, they held nothing back, sending math their scathing reviews and imploring math to either leave them alone or welcome them with open arms. The students discussed math as they related to it, and it is only in this type of deep reflection that I could start making sense of this relationship and thus create goals for how I might take part in that relationship in a deep and meaningful way.

The writing and reading of Dear Math letters is worldbuilding. It is a process of centering student voices in the work to continually improve math classes to meet their needs. I fell short a ton. Some students didn't like my methodologies. They would get frustrated that my math class felt different. Some said it felt more like therapy than a math class. And others couldn't get past the notion of math as an answer-getting, performative subject. Still others had received lots of glory in previous math classes and were disoriented to be in a space where they weren't continuing to receive these same types of celebrations in the form of grades.

I hope that any of those frustrated students will acknowledge that I sat with them in their frustration and worked with them to co-create the world they sought. Similarly, I hope that as you read this book, it will be worldbuilding—that it will provide you with space to exhale, dream, laugh, lament, and imagine a world of math class that centers on students' storytelling and experiences in the quest for a more whole, human-centered, beautiful way of teaching.

CHAPTER 1

DEAR MATH, YOU ARE DREADFUL

"Dear Math, I hate you! You fill me with deep dread in my everyday life."

ELI WALKED INTO class on the first day of the semester and slumped down at the table where his name was written on a three-by-five card. On the screen, instructions asked the students to add information to the card, namely, preferred pronouns, interests, and one mathematical strength. By the end of the three-minute timer, Eli's card was empty and his head was down on the desk. I knelt down by his desk.

"Well, you haven't really given it a chance," I claimed, attempting to infuse lightness and truth into the young man who was clearly grumpy. It was, after all, the first day of class.

"Oh, I've been doing this for twelve years. I hate math, and there is no reason for me to learn it. I don't see why I would give it a chance," he lauded.

I took a deep breath and looked at the other twenty-four kids in the room who were ready for learning and my attention. Their body language was in stark contrast to Eli's, which was seeping with dread.

Knowing I wouldn't win over this student in a day, I said, "Okay, I hear you. I hope to hear more about what brought you to this point of hatred for mathematics. For now, can you at least agree to share your thinking on the puzzle we are working on today with our group? I know your group will appreciate what you have to say."

He rolled his eyes, and I walked away and launched into teaching the class. I knew the next day, he would be taking on my Dear Math assignment, and I would get to hear more about the journey that led him to this point.

The next day, he wrote ...

"Dear Math, You are boring, repetitive, and not interesting. You always are difficult. You will never understand. Every single mistake causes even more problems. Math, you have always been my least favorite class. I never got the chance to fully understand the different math concepts, which would make me frustrated and not care to continue to struggle. It made my brain hurt. Although I truly dislike math, I can say it has been one of the more challenging obstacles in my life. I would avoid asking for help to try to learn it myself, but I continued to struggle. This has made me think in different ways and made me work a little harder to think outside the box. Although I can keep talking about my frustrations with math, one way math has helped me grow is by allowing me to be more patient because I know it will take time to get the reward, in this case, the answer. One thing I do look forward to in the future is increasing and improving my ability to solve math problems by myself. I also think that math should not be rushed. So many times, I hear students saying they aren't done or need more time, and I thought we were supposed to work at our own pace. So I think it should be structured to put the groups of kids who are in the same understanding of the topic. From, Eli."

SOURCES OF DREAD

Dread is an adjective meaning "to involve great suffering, fear, or unhappiness; extremely disagreeable." While many students don't name dread explicitly in their letters, other words arise, like hatred, cruel, pain, turning clear skies gray, unenjoyable, and annoying. We selected the term "dreadful" to sum up these letters because it seemed to truly capture the depth of dislike this group of students had for math.

> "Dear Math, Why do you even EXIST? I'm genuinely asking because it is such a complex subject. I don't want to be rude or anything, but you can be a PAIN sometimes to everyone. I've had other people agree with me on that, but it's something we can overcome." —Edwin, tenth grade

Why do so many students find math class dreadful? And why do so many adults have negative comments about the subject?

For Eli, his dread came from math being "repetitive, boring, and not interesting." For Edwin, the dread came out of it being "a complex subject" and included questioning the purpose of math's existence. For Hayley (whose letter appears later in this chapter), the dread stemmed from a feeling of being "dragged along" without understanding, while math itself was holding on to so much certainty. For other students, the dread came from a bad experience, a poor grade, the stifling of their thinking, or the way they were treated in class if they didn't know something.

While the stories in the letters are powerfully descriptive, the statistics are alarming. In some school districts, over 50 percent of students fail their freshman math class. The situation doesn't improve in college. Rather, a 2011 report from the Education Commission of the States noted that many researchers believe the failure rates of college algebra classes nationwide to be around

50 percent. It's no wonder that widespread dread exists for being required to do something that often leads to failure.

"Dear Math, You are best classified as a regrettable necessity." —Andrew, twelfth grade

It seems students and adults of all ages dread math, so when does it start? I have worked with teachers and led an exercise in storytelling about their math experiences. One activity we do is create graphing stories. Many teachers describe their learning journeys with math like the examples in Image 1.1.

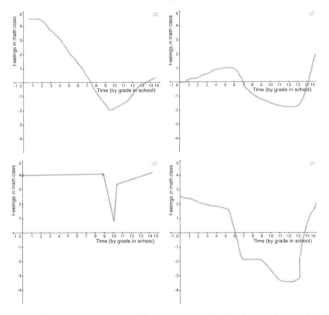

Image 1.1: Graphing stories of feelings in math class by grade in school.

Students express positive feelings in early elementary school, and then by middle school, a negative trend begins. Many of their story-lines head back upward if they begin teaching math, but those negative years are extremely common. In students' Dear Math letters, many of them identified their "moment of fracture" in their math identity as occurring in third or fourth grade, while the second-largest percentage was in middle school. Further research on math anxiety shows that it might not start at one age in particular, but it was always accompanied by some type of test or performance (Boaler 2017).

"Dear Math, Never liked you, never will, you cause me so much pain." —Rayon, seventh grade

The feeling of dread for mathematics may be classified as stemming from pervasive math anxiety—a problem so massive that a whole field of research is now dedicated to it. A study conducted in 1978 concluded that about 68 percent of students enrolled in mathematics classes experience high mathematics anxiety (Betz 1978). More recently, a 2009 study showed that about 17 percent of the population have high levels of math anxiety (Ashcraft and Moore 2009). Societally, we perpetuate this thinking, normalizing the idea that not being mathematical is just a regular consequence of life. People joke and make light of this type of thinking, whereas the idea of being illiterate is much less culturally acceptable and rarely the butt of jokes in the media.

"Dear Math, I hate you; you make my clear skies feel gray. In a world without you, I don't know what I would do, though your own significance doesn't have to involve me. I've never liked your certainty of 'right' or 'wrong' when it ends up with just me unknowingly being dragged along." —Hayley, twelfth grade

Because math invokes such strong emotions, often feelings associated with dislike and dread, I hope that we can hold space for these emotions and create activities where students can explicitly share their stories and unpack their feelings. Beyond caring about math and our students, we need to care about our math stories, particularly mathematical identities: how students see themselves as mathematicians and participate in mathematical spaces.

When he was president of the National Council of Teachers of Mathematics, Robert Berry stated, "Effective teachers affirm positive mathematical identities among all of their students, especially students of color." But why should we care about our students' mathematical identities? Can't we just teach them the processes they need to know?

The answer to this is, as we will see throughout this book, an emphatic "No."

The learning process inherently includes the development of identity. In his book on communities of learners, Étienne Wenger (1998) explains, "Because learning transforms who we are and what we can do, it is an experience of identity. It is not just an accumulation of skills and information but a process of becoming—to become a certain person or, conversely, to avoid becoming a certain person. Even the learning that we do entirely by ourselves contributes to making us into a specific kind of person. We accumulate skills and information, not in the abstract as ends in themselves, but in the service of an identity."

Students naturally form their own mathematical identities with or without our involvement. If we want students to engage in a better relationship with math, we need to guide them toward more positive mathematical identities. Recognizing how the unpleasant feelings they already have affect their identities is a strong first step.

"Dear Math, You are a cruel, heartless mistress." —Tony, twelfth grade

Being able to share such strong emotions clearly creates the space to forge a path out of these emotions. Students may not even need advice or solutions for the problems they are experiencing; they may just periodically need to vent. Students come into our classrooms each day with a great variety of stories. If we do not create space for them to share their stories, then we are making it more difficult to help them create healthier relationships with math overall.

A LESS DREADFUL EXPERIENCE

Dear Math letters are a critical tool for understanding and overcoming dread for two related reasons. The first reason is that the letters give students a space to share their story, vent, and unpack the ways they have become the mathematician they are today. We can normalize experiences from the past, process them, and collectively make sense of a path forward.

The second reason comes from the teaching standpoint. If we don't ask, then we are designing curricula and making instructional decisions by relying on our assumptions from prior experiences, our own math experiences, or feedback we get from the loudest students. I used to follow the ignorance-is-bliss concept, ignoring how my students already felt in favor of making my class as awesome as possible to help them love math. How wrong I was. I regularly assumed that the students were as ready to think about math as I was and that they were excited to learn it in the same ways that had excited me.

I am reminded of Chimamanda Adichie's famous 2009 TED Talk entitled "The Danger of a Single Story." In it, she states, "The single story creates stereotypes, and the problem with stereotypes is not that they are untrue but that they are incomplete. They make one story become the only story."

"Dear Math, I have always hated you; I can never do you. Sometimes I get the answer, but that's only on the better types of math. But I guess I need you because those 'sometimes' are important in daily life. But I still hate you. I never look forward to doing you, I always look forward to finishing you and going on with my day." – Sam, tenth grade

Mathematics classrooms are easy spaces to become "one story." There is a math problem, there is a way to solve it, everyone tries it and does well or doesn't, and then we move on to the next problem. Dear Math letters hold space for and give voice to all the different math stories in the room. They allow healing for those with traumatic math stories and encourage the co-creation of stories that are whole and complete. Most importantly, they tell us things we wouldn't have known if we hadn't asked.

Every time I open a new Google Drive folder containing my students' Dear Math letters, my heart starts beating a little faster. I know some letters will reveal a dread for the subject that I must engage students in each day for an entire semester. I'll have to address their feelings in the ways that I teach. If the dread that students feel is connected to feeling rushed, we might try out fewer activities in our class that focus on speed (Chapter 6 has great examples of this). If their dread is connected to their grades, we might consider alternative grading activities or more equity-oriented assessment strategies (Chapter 5 has ideas for this).

But if I didn't ask, then I might delude myself into thinking that everyone was walking into my class ready to have a good time. Furthermore, I might unintentionally impose my own math story and identity markers onto my students, joining the oppressive figures I sought to dispel. Delusion and oppression are unhealthy starting places for a semester of work together.

DEAR MATH: HOW TO GET STUDENTS TO START WRITING

So, how do we engage students in the Dear Math letter writing? The first step is to necessitate the letter. Oftentimes, students aren't used to writing in math class. They also are unfamiliar with being reflective and metacognitive about the discipline of math. Lastly, they likely haven't ever personified math and "talked to it."

I usually set up the assignment by saying:

"As we start this school year, it is important that I get to know you as mathematicians. Who you are as mathematicians and the stories that formed you before you got here are incredibly important to me. It is my goal that you would all feel like you are learning and growing each day as individuals and as a collective math community. Without understanding who you are, I cannot properly design and facilitate this math classroom for you. I'll start by reading my Dear Math letter to you.

After my story, I ask, "What did you hear in this letter? What surprised you? What parts of my story relate to yours?"

After the discussion, I say, "Today, we are going to write 'Dear Math' letters. I look forward to learning more about your math story. Thank you for sharing your stories with me."

Questions for Prompting:

1. Tell me about a time in elementary school when you felt successful in math class. What happened?

2. Tell me about a time in elementary school when you struggled in math class. What happened?

3. When your friends talk about math, what do they say or do?

4. What is one way that math has helped you grow?

5. What is one of your greatest mathematical strengths?

6. What is one of your greatest mathematical challenges?

7. How do you plan to engage with math in the future? (Going into a STEM field? Using math in your career? In your life? Tackling complex problems in a systematic way?)

8. What can you thank math for?

9. How would you change math classrooms?

10. What would you like more of in math classes?

PROCESSING DEAR MATH LETTERS

After students submit their letters, I set aside about an hour per class to read over the letters and listen to students' stories. I underline, share celebrations and connections, and thank them for sharing their letters with me. While it is easy and natural to feel defensive and jump into "solutionitis" when reading the letters, particularly ones that are steeped in dread, I try to receive the stories just as they are. As we will see in the coming chapters, the work of forging a new identity will happen during the semester and year, not just in the moment of writing and responding. This first pass at the letters gives me valuable narrative data. This data comes in extreme stories of dread (like the ones referenced in this chapter), extreme expectations for the class that feel out of line with my hope for the class (like the letter from Eli, who asks for academic tracking, a practice that is out of alignment with the philosophy and design of my school), and overall trends.

A variety of stories should be expected from a class, and if there is a disproportionate amount of negativity or dread in a classroom, then it is imperative that you create specific activities to acknowledge and transform on a class level. In this case, I will often share some student writing (anonymously) with the class and then share a goal and track progress toward that goal. For example, I might say, "My goal is that you find value in coming to math class each day, and I'm going to design activities that help the lessons feel valuable to you. Each week, I will give you an exit ticket, and I will ask you if what we did this week felt valuable. In this way, we can keep getting better at meeting our goals."

PROCESSING DEAR MATH LETTERS WITH STUDENTS

Another activity I have done with the Dear Math letters is to have students exchange their letters and read the other person's letters for connections and surprises in the stories. This could be anonymous or, if expectations are stated from the beginning, with the names intact. When students read each other's letters, they immediately begin an empathy-building exercise that serves both to understand that your math story is just one of many and that our math community is composed of a beautiful, diverse set of learners who all hold value in the space.

I have had students read over their own Dear Math letters with different lenses depending on our work that week. In one project, we were working on understanding our shifting math identities, and the students needed to self-identify a "fracture" in their math identities in their stories. After each student identified a "fracture" and the grade they had been in when it happened, one of their assignments for the remainder of the semester was to find a student in that grade and offer them support on their math homework. The students who did this assignment attested to feeling great about helping other students and about realizing that the types of math that had caused such deep wounds in their math identities were now very doable for them.

Dear Math letters are certainly not the only way to promote this type of storytelling and identity unpacking. For more activities and projects to help unpack students' mathematical identities, refer to Appendix A: More on Math Identity.

THE IMPORTANCE OF MATHEMATICAL IDENTITY WORK

As learning theorist Yrjo Engestrom (1995) stated, "Identity work is never 'done,' it is always ongoing. Although a person's identity is not determinable, neither is the meaning-making involved in identity work entirely free but, instead, is mediated by the discourse and practices of people's communal social activity systems." Because of this, we create space for students to share the stories that formed them and for the possibility of evolution in those stories over the course of the year. The possibility of evolving is related to the idea of a growth mindset, and, while it's not the only point, believing that success can be found is an important step. Even day to day, the ways students feel about themselves as mathematicians can shift dramatically, but we can design a class where they can flourish when we tune our eyes and ears to their stories and ways of being in a math class.

> "Dear Math, I have hated math ever since third grade; it's annoying and unenjoyable. It used to be that I liked math, but that all changed in third grade when we had to learn our times tables, and I was always stressing. I like normal multiplication, the kind where you can ACTUALLY take your time, but not this." –Andrea, seventh grade

OVERCOMING DREAD IN THE CLASSROOM

Isabela and I met when I was her teacher in her freshman year. As a student, she seemed driven and justice-oriented. As a mathematician, she was brilliant at organizing information and she asked many questions, yet she lacked confidence. One of the first times we met, she told me that she had test anxiety, and as we worked together, I noticed that her anxiety was pervasive in her work. She would rush to an answer, second-guess her thinking, and then her brain would "shut off" (her words), and her emotions would take over. In her sophomore year, she wrote a Dear Math letter in which she unpacked this anxiety and the resulting feeling of dread that was now a part of her heading to math class. Her letter that year read:

> "I really like you. But you don't come naturally to me. I have to work extra hard to understand and really

conceptualize what you have to offer. There have been times where I have felt discouraged, frustrated, and exasperated, especially on tests, which is where I believe I can never fully express all of the things I know in a way that helps me be successful." —Isabela, ninth grade

By reading and responding to her Dear Math letter and giving her space to unpack her story and mathematical identity, Isabela's teachers were able to dig deep into what was blocking her achievement and connections, and they highlighted her strengths. From there, they helped Isabela build a new story for herself about who she was as a mathematician.

I had the opportunity to teach Isabela again her senior year, and, as we always do, Isabela wrote another Dear Math letter, reflecting on her mindset growth and identity during her high school experience. She wrote:

"While the term 'math growth' might inherently imply academic growth, I think for me it's a lot more about a shift in attitude and my reactions when I am faced with challenges. I developed a sense of patience and open-mindedness for the first time ever. I no longer got as frustrated with myself when I didn't understand something and would allow myself to take my time. As I reflect on my past experiences and emotions related to math, I can confidently say that I have a strong foundation. And this is a great amount of growth for me because two years ago when I wrote this letter as a sophomore, I could not say that I felt like I had a strong foundation in math." —Isabela, twelfth grade

GIGI'S REFLECTION

The notion that math is dreadful is not a terribly uncommon one. This dreadful relationship depicted in these student letters is a mere echo of the greater public opinion. In fact, it's so common that it has become a trope in the media. My favorite example is in the movie *Mean Girls*, when Damian memorably embodies dread upon hearing Cady's plans to join the Mathletes and fearfully notes, "That's social suicide!"

Damian's dread perfectly represents the cultural conundrum of math: if you're bad, you're stupid; if you're good, you're a nerd. I can recall feeling the effects of this mathematical fork in the road, becoming stuck shooting for some nonexistent gray area or third path. In the early days of my freshman year, the anxiety of "the decision" crept toward me, appearing on whiteboards and through hallways. Which path would I choose: "scholastic slowness" or "social suicide"? Although it may seem as though the ultimate moral lesson in every fable, story, and movie is to rise above the social pressures of coolness and run into math's comforting, outstretched arms (Cady does end up joining the Mathletes and learning incredibly pertinent lessons wildly applicable to her situations in the greater world—how perfect!), doing so is rarely so simple or perfect. Both paths, in this case, are surrounded by pressure and labels. Eventually, I made my choice not based on what I thought would be the best, most academically and socially fulfilling route for me but on the outside pressures in my life. My family's aspirations for me just outweighed my classmates' opinions, and so my fate was sealed (time to buy up all the fanny packs and satirically large glasses in town): I was a nerd.

This decision brought about a whole other type of dread, one that pressured me to constantly act at peak performance. To get a wrong answer, especially publicly, was to move backward on the path. This pressure was exhausting and caused a lacking math mindset. When you're too afraid to be wrong, you end up missing out on learning why you aren't right. While this demand for constant correctness did admittedly result in a relatively high GPA, it was rendered arbitrary in terms of genuine learning. Ensuring my report card remained devoid of any number beginning with a three was exhausting and all-consuming, leaving me without sufficient room to equally value my grades and my understanding of the concepts being taught.

Throughout my first semester of ninth grade, I learned that the notion of the binary paths of arithmetic, much like Cady's limit, "doesn't exist" and, furthermore, that sometimes there is no moving backward or even forward. I learned that math is merely (and beautifully) the conceptualization of answering questions. It brings tangibles to the intangible. Math is one big, beautiful mess, like a game of Chutes and Ladders, which allows for high levels of performance and true understanding. In studying it, you might end up going up, down, left, or right, but no matter what, you'll always be learning. That doesn't seem very dreadful at all.

INVITATION TO

REFLECT

- How does your math story shape how you teach math?

- Why might it be important to recognize our students' math stories and math identities as we design our classrooms?

CHAPTER 2

DEAR MATH,
YOU ARE HIERARCHICAL

"Dear Math, All of my friends understood you when I didn't."

AFTER I ASKED my class a warm-up question one day, Zelma stated loudly and with certainty, "I've never been good at math. I don't know anything about this." I looked at her with practiced curiosity in my eyes, avoiding my inclination to correct her deficit language and then look around the room to make sure the peers who heard also caught my silent correction— that in our class, we don't talk about ourselves that way.

On the one hand, I appreciated her bravery and vulnerability. She was ready and willing to admit that she was struggling to understand our current work. On the other hand, I wholeheartedly disagreed with her self-assessment. She was a brilliant performer and questioner. She was loved by her peers and regularly made people laugh. She was wildly independent, a creative thinker—and struggled to turn in most of her work in a timely fashion, thereby getting low grades in most of her math classes.

As I listened to Zelma talk about how bad she was at math, I became focused on the reference points for her claim. Her grades, test scores, and feelings were all part of her story, but the main part I heard was her understanding of herself in relation to her peers.

"All the other kids are fast and get their answers right. They know all the confusing formulas. It's like they just have them in their head."

"Do you know how your peers view you?" I asked.

"Probably like I'm stupid," she responded.

I was saddened by this evaluation. Sure, Zelma was known by her peers as a work-avoider, and a few of them probably saw her as less than helpful in group projects, but I felt certain that nobody saw her as stupid. They saw her lead class discussions, a task that was challenging for many of them. They saw her ask questions when nobody else would. And they saw her sometimes stay after class to figure out a particularly challenging problem.

I got to thinking: How can I foster a sense of community in my classroom where ALL students are known to be integral parts of it? How can Zelma acknowledge her differences from her peers but know that they are appreciated, good, and necessary? What role do I play in the creation of this environment? What role does Zelma play? What roles do her peers play?

Zelma was an essential part of our classroom community. I set out to explore ways to highlight her strengths amid an academic world that diminishes her because she doesn't fit the specific ideal of a "successful student."

LET'S TALK ABOUT HIERARCHY

Mathematically, hierarchy is a word that refers to a predetermined order of a set. The natural or counting numbers, for example, necessarily follow an order where two comes after one, and so on. This same notion of hierarchy (coming ahead of and behind in a particular order) is pervasive in student language as they write letters to math. In the letters that kick off this project, we read words like slow, behind, good enough, keep up, and more. Though no students explicitly used the term "hierarchy," it sums up a big theme across letters where students were deciding their mathematical worth and defining their math identity based on comparisons to their *perceptions* of their peers.

"Dear Math, When others talk about you, I feel like they know you better than me." —Mandy, tenth grade

Students' language in Dear Math letters regularly references this sense of social comparison and ordered thinking as critical elements of a math identity. In this chapter, we explore how assumptions are made about ordered learning and students' progress on an ordered path. We explore the ways students tell stories about how they fit into a perceived order and how that ultimately describes them as a mathematician. Although this practice of ordering and ranking ourselves is not unique to math classes, we find that using it in math classrooms plays strongly into students' feelings of belongingness and self-worth because of the undue power that math holds in society.

In some student letters and in our follow-up student advisory panels on the topic of hierarchy, students describe this hierarchy as a pyramid (see Image 2.1). The teacher is often at the top of the pyramid because the teacher knows all the information and is expected to provide it to the students as needed. After the teacher at the top, the next level shows the "smartest" students in the class, or the students who seem to know the most information and get the best grades, as perceived by their peers or as lauded by the teacher. The ordering continues to the bottom of the pyramid, where students express their desire not to "be the worst." My advisory group of student editors on this chapter noted insightfully, "There is no comfortable spot on the pyramid because the people at the top are stressed out trying to make sure they keep their spot, and the people at the bottom are busy feeling bad about themselves at the bottom."

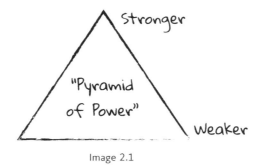

Image 2.1

Another way of thinking about the hierarchy is as a "narrow path." Amy Noelle Parks, a scholar in the hierarchy of mathematics education discourse, wrote about "a narrow path" metaphor by which students are described as heading in one direction on a path, and at any given time, they are at a fixed point on the path. Necessarily following is that some students are ahead on the path and other students are behind. Students share findings congruent to this analysis as we read in their Dear Math letters that they use language like "ahead" and "behind" or "slow" and "fast." Parks goes on in her work to state that this metaphor of a narrow path that students are on is pervasive in classrooms, schools, teacher ed programs, and even curricula. The language that is used to describe students is almost exclusively a linear learning journey (Parks 2010).

"Dear Math, I'm sorry for despising you ... All of my friends understood you when I didn't, and that hurt me. It made me feel insecure, like I wasn't good enough. That I wasn't worthy of your knowledge, that I wasn't smart enough." —Osha, tenth grade

The analogy extrapolates to describe problems when students aren't where they "should be" on the path. If a student is behind where they "should be," they might need to be in a different class or receive remediation. And if they are ahead of where they "should be," they are advanced. Many teachers even design their entire classrooms around this notion of ahead and behind and, instead of designing lessons that are appropriate to the standards of the grade they are in and then building in support for students who need more tools to access the curriculum, they end up teaching to students' perceived deficits.

A hypothetical teacher quote for this type of scenario might be: "Wow, *none* of my eleventh graders knows fractions? I guess I'd better spend a few days walking back down this linear path to Fraction Land and review it for them. Certainly, when I explain it to them, it will make sense once and for all!"

How prideful and delusional we secondary math teachers can be, getting to "blame all the teachers who came before" for our current students' placement on this narrow path of doom. In mathematics classes, the notion that students can compare themselves hierarchically frequently stems from our assessment and grading practices, teacher language, and curricula.

> "Dear Math, I used to never get math; it was always too difficult. I thought that I would never understand it. I really want to learn math more so I can understand it easily. I need to learn math one way or another. I always hear from everyone that people learn things at their own pace, but sometimes I think that I'll never be good at it. I want to be able to understand math for real-world problems." –Kamryn, seventh grade

Similarly, Rachel Lambert, a math education researcher, speaks of the ways we talk about students in our math classes. She addresses the myth of high kids and low kids, likening the progress through math as a winding staircase heading in lots of directions (think M.C. Escher paintings), as opposed to one straight ascending staircase. She asks, "Why do we talk about kids like this? We are trained to rank and sort kids into groups, but it doesn't work."

She says that thinking about kids like this starts to change the way we view our classes and plan our lessons. Much like the earlier example, we start to plan remediation lessons for some and enrichment for others. She concludes by stating that "Kids aren't low: our expectations are."

"Dear Math, My current feelings about math are, 'This is really confusing because everything stopped making sense after mid-sixth grade. Math got really annoying because I couldn't keep up like everyone else when they seemed to get it at the same time as each other." –Theo, seventh grade

The National Council of Teachers of Mathematics recommends that schools offer rigorous and integrated math classes for all students in a pathway through grade eleven, and then midway through that year, students can branch off into various pathways per their hopes for college. This is in stark contrast to schools that start separating kids as early as elementary grades based on their perceived abilities on a usually small set of indicators. The damage is done not only to kids who have been deemed as belonging on the "low" path but also to kids who are placed on the "high" track, who thus miss out on the opportunity to learn a variety of ideas and perhaps develop a fraudulent fondness for mathematics (more on this in Chapter 6).

However, all of these metaphors are insufficient to describe the community we strive for in our math classes.

DISRUPT HIERARCHICAL THINKING WITH BELONGINGNESS

Disrupting the narrow path and pyramid metaphors that students use to describe themselves is no small task. Hierarchical and ordered language is pervasive in all parts of our capitalistic society. While this language is helpful in making sense of systems at times, it is important to step outside of that framework, particularly when it comes to our ways of relating to each other in math class.

If the narrow path metaphor relies on a limited view of mathematics, ourselves, and others, I propose that a disruption to this metaphor begins in close relationship with others in our community and celebrates the strengths and mathematically creative solutions that each person brings to that community. Where hierarchy seeks to order and separate, belongingness seeks to diversify and connect. Belongingness is defined by Ilana Horn, associate professor of math at Vanderbilt University's Peabody College, as "students experiencing frequent pleasant interactions with the sense that others are concerned about who they are and for their well-being." She states that classrooms centering on belongingness must hold "clear norms and expectations about how students treat each other."

> "Dear Math, Another thing I'm angry about you is that I never saw anybody else feel the same way about you. I always saw the other kids just easily go along with your weird complicatedness, but I couldn't." —Andrew, tenth grade

If a student feels belongingness and they sense that they are an important member and their presence is valued daily in their community, they become less likely to compare themselves to their peers and remove themselves from the possibility of success in the classroom. Connecting daily in a variety of ways gives students multiple opportunities to enter into groupwork and classwork situations with agency and wholeness.

Research on belongingness cites the importance of explicitly designing opportunities for it into daily routines and practices. While curricula may play a role in belongingness, it is the teacher, as the holder of authority in the classroom, who has the power to create a space where students feel belongingness both from the teacher and from each other.

In their non-cognitive report, the University of Chicago Consortium on Chicago Schools Research noted that "When a student feels a sense of belonging in a classroom community, believes that effort will increase ability and competence, believes that success is possible and within his or her control, and sees schoolwork as interesting or relevant to his or her life, the student is much more likely to persist in academic tasks despite setbacks and to exhibit the kinds of academic behaviors that lead to learning and school success. Conversely, when students feel as though they do not belong, are not smart enough, will not be able to succeed, or cannot find relevance in the work at hand, they are much more likely to give up and withdraw from academic work."

"Dear Math, Why do you like other people but not me? I guess I don't have the math gene." —Julian, tenth grade

We must attend to the notion that students should think, "I belong in this academic community." Learning is a social activity, and it is frequently constructed through interactions with peers. As such, it is important for students to know they are part of a community and that they are invited to be their whole, true selves in classroom spaces.

BELONGINGNESS BUDDIES

I walked into Carlee's classroom for our second observation of the year. I was her mentor, but we were continually learning from one another. She had arranged the students in groups of four, and they were documenting their phone numbers and Instagram handles for one another. They were to be belongingness buddies for the next four weeks. As BBs, they would check in on each other in the evenings after the school day, they would sit together in class, they would collaborate on challenging problems, and they would come to know one another's strengths and areas of growth as mathematicians. In short, they created a mini-community in the larger classroom community. Although the check-ins were largely content-based, Carlee often asked the students to check in on each other's feelings as well. This practice led to a wonderful sense of community in the classroom, and even more so, a feeling that everyone was an important part of the space in equal and different ways. One way of contributing was not elevated over another, and only in getting to know each other well were the students able to truly recognize the value that each of them brought to the classroom.

I started using belongingness buddies in my classroom in 2018 and continued to use them through the pandemic and distance learning in 2021. I modified the practice slightly each year, but the underlying message remained the same: You (all of you!) are integral members of this classroom community, and we cannot function at our best when we aren't all present physically, mentally, and emotionally.

The first thing belongingness buddies attended to was the actual physical presence of each other. When you arrived at class, were your belongingness buddies present? If not, did you text them and check in on them? Did you offer to fill them in after class? Do they have something important going on that the teacher needs to know about? This was a general part of our warm-up routine. As the students brainstormed about a prompt on the screen, I wandered the room, noted empty seats, and prompted the students to reach out to one another. Some students who were routinely absent would become my belongingness buddies as a secondary step. As the class gathered, I would shoot a quick text message to the missing students and just check in on where they were.

Often, students would report messages like "Lou is on his way; he just has a flat tire," or "Lourde is sick today. I'm going to call her later and get her what she needs!"

After a couple of weeks of my bringing up belongingness buddies at the beginning of class and prompting text messages, the students started doing it without me reminding them.

As teachers, keeping track of absent students is a challenge. If we can leverage the groups to keep track of their members, we are helping not only ourselves but the whole community. They start to notice one another in both social and academic ways.

The second activity of belongingness buddies is to share math reflections at the start and end of a week. They might share some goals for the week and check in on those goals at the end of the week, or they might be accountability partners for work completion on our current portfolio or project. They regularly celebrate each other's work in small groups and to the class out loud, and they give feedback on work that needs to be revised.

While not specific to math class, the practice of belongingness buddies sets a tone for collaboration and community that is much needed if we are to create spaces of belongingness and communities that disrupt hierarchical thinking in our classes.

BY ANY OTHER NAME

Admittedly, the term "belongingness buddy" might come off as elementary. My classroom vibe was such that I could use the silly tone pretty handily. If it's not your thing, try out a different name, like check-in partner, collaborator, or weekly partner. The name doesn't matter as long as the purpose is clear: students are looking out for each other and making clear attempts to connect with, give feedback to, and celebrate each other.

BELONGINGNESS BUDDY WARM-UPS VIA VIDEO CHAT

In Fall 2019, one of my students had to leave for a week for an athletic competition. She was trying to keep up with her work while she was gone, but she knew she would miss our "Daily Discourse" warm-ups that were an integral part of our classroom environment (see Chapter 4 for more on this routine). She and her belongingness buddy agreed that they would FaceTime each day during the discourse so she wouldn't miss it. Her buddy called her and placed the phone on her desk so she could listen to the conversation and even contribute on occasion. I appreciated the pair's prioritization of the warm-up. Even though one student was on a trip, these students worked together to sustain the community.

On another occasion, a student shared with me one morning, "Ms. Sarah, have you heard from Will lately? I have tried to reach out three days in a row, and I haven't heard from him. I called and texted and checked Instagram, and no luck. I'm starting to get worried." I thanked this student for her concern and realized that I, too, hadn't heard from Will in three days. Upon calling home, we learned that Will had had a falling out with their family and had run away to another state. We were able to get in touch with the student and find a way for them to continue their learning virtually while also connecting them to helpful services. We reached Will quickly because a student was looking out for their buddy in the community.

On the wall of my office, I display a quote by one of my mentors and the founder of my school, Rob Riordan. It reads, "Be the one who notices." As teachers, we have opportunities to be that person who notices and sits with students in their moments of need. We notice the student who is sitting quietly in the corner and the student who is crying out loudly to be heard. We notice the sixth grader who can't seem to sit still and the high school senior who can't decide whether college is the best option for them. But it can feel impossible to notice everyone's needs all the time. Having belongingness buddies equips students to "be the one who notices." As we spread out the noticing, we create communal spaces of care, advocacy, and respect, disrupting the hierarchical thinking that seeks to subvert it all.

For a resource on belongingness buddy routines, see Appendix B.

BELONGINGNESS IN THE CLASSROOM

It was exhibition day for my class of seniors during the fall semester while distance learning during the pandemic. In preparation for the day, the students had completed posters using functions as a storytelling mechanism for their neighborhoods. They had gone through critique and revision cycles, and by that day, they were to have their work in final draft form.

As is always the case in a group of learners, the final products were in various stages of "done." Nevertheless, we persisted with the exhibition, acknowledging that we still learn a lot even when displaying only partially completed work. We also learn a lot when we participate in others' presentations and see new ways of improving and making connections in our own work!

In one breakout room, a student named Kal had created a beautifully polished piece of work and had gone through a couple of revision cycles to get it that way. It was clear that he had spent many hours on the project in preparation for exhibition day. Admittedly, Kal would be considered by his peers to be a strong mathematician. The two other students in the room had also completed work, albeit, according to our work quality rubric, with less precision. It was also clear to me that these students had spent significantly less time working on their projects.

I popped into the room to listen in on the share-outs. Kal, the student with the highest quality work, would definitely have the perceived power in the room. He carefully shared his findings with the group, and the group members applauded his effort and looked on in awe about what he had created.

"Mine will never be as good as yours," Sandra shared.

"Of course, it will," Kal stated. "You already have a clear function and really interesting storytelling elements. In fact, I like the way you organized yours better than I organized mine."

"Yeah, but the math is just easy for you. It's really hard for me," Sandra responded in self-doubt.

"Math is really hard for me too. I'm a really slow thinker," Kal responded.

I watched Sandra's jaw drop. It had never occurred to her that Kal worked hard to be the mathematician that he was. She thought he was just good at things, and by comparison, she was just bad at things.

In this scenario, Kal did something I don't often see in classes (or in society)—a student used their power to lift another up. Given that the perceptions of hierarchy are strong, the best we can do is try to disrupt them and support our students in this disruption. It is only through these coordinated efforts that we can forge a new relationship outside the hierarchy. In sharing his story and simultaneously validating Sandra's work, Kal disrupted the hierarchy, helped Sandra feel belongingness, and supported her in charting a new path in her math development.

The practice of math can help us grow as people. As author Francis Su said, "You love all people when you stop speaking about talent as a thing either you have or don't and start speaking of virtues that each person can build through hope and the joy of

45

courageous effort and hard work. To love is to believe that anyone can flourish in mathematics."

GIGI'S REFLECTION

The "dreadful" descriptor of math is obviously common among young people. Everyone knows of the "math sucks because it's hard" mindset, but the fear of being wrong stems from the lesser-known source of math contempt: the mathematical hierarchy.

It seems inherent that we assign value to one's speed and accuracy in math. We can't seem to stop doing it. It's like drinking soda. We *know* it isn't good for us. We *know* that there are better options. Yet time and time again, we keep on reaching for that cola. I note this even as an individual who has attended de-tracked, project-based learning schools for thirteen years. Even without the element of tracked classes, that pressure to complete the test or worksheet the fastest is still present. This pressure creates a skewed, one-road-to-correctness lens, which creates competition, which creates winners, which creates a hierarchy. It's an all-too-sad, all-too-real, and all-too-common path to students' viewing math as classroom warfare—and to their subsequent capitulation.

Take timed multiplication tables, for instance (note: don't worry, that chill you just got down your spine from merely thinking about the nightmare that was this math "exercise" happens to all of us). The kids who spend time analyzing the problem set, using their practiced tools, and tangibly drawing out the math are punished. The kids who do it all in their heads and finish first are rewarded. Just like that, a mathematical oligarchy has taken over the classroom, thereby destroying the learning environment. From that point forward, those kids who worked, tried, struggled, and retried are labeled as less than, sporting a scarlet C+. The students who finished quickly or correctly group up and work together, as do the others. This means that neither group can properly learn or grow from the other.

This is where belongingness buddies come into play. In my ninth grade year, Sarah introduced belongingness buddies: a small group of people who meet every day, discuss math, check in on progress, and so on. At first, I was upset that I couldn't work with my friends, the same group I had been doing math with for years. I sat down at my table with my randomly assigned belongingness buddies and felt very uncomfortable. Math, to me, had previously felt like such a vulnerable thing, something I only wanted to do with those I trusted most. Not to mention, we all worked at wildly different speeds, fostering some frustration when one would have to purposefully slow down or frantically speed up. Eventually, though, we began to find our rhythm. Together, we figured out the different strengths each of us brought to the table. Together, we became more comfortable around each other. And together, we worked through math with a communal and genuine understanding.

Through these belongingness buddy meetings, whether they were for five or forty-five minutes a day, we slowly and tentatively subverted that hierarchy. Our meetings impacted me and those around me, forcing us to unlearn the practice of putting one skill or one student above any other. Practices such as this one, which purposefully disrupts pre-existing separations or hierarchies, are integral in the process of preserving mathematical creativity and love.

INVITATION TO
REFLECT

- What language have you heard students use that reflects a hierarchical way of thinking about learning mathematics?

- What classroom moves might you make to disrupt hierarchical thinking and build a community of belonging in your classroom?

CHAPTER 3

DEAR MATH,
YOU ARE UNNECESSARY

"Dear Math, Maybe I'll try to find a way around you."

AS I LOOKED over responses on the exit tickets for the week, the words "Why are we learning this?" pierced my soul. My students and I had been exploring a unit on logarithms as part of our eleventh grade curriculum. Even though I sincerely doubted that many of the juniors sitting in front of me would actually calculate logarithms in their lives, I felt beholden to do them justice in our eleventh grade class. Our school isn't super-rigid in our use of curriculum, but our "suggested scope and sequence" has strong suggestions for what students should be thinking about at each grade level.

When a student asks me, "What is the purpose of studying logarithms?" possible responses swim around in my brain, including:

1. Well, logarithms are used in pH levels, so anyone in the science field will need them to …

2. Let me pull out my trusty "when will I need to use this" poster that was shared in my teacher prep

program and go to the line labeled "logarithms;" certainly, we will find the answer to your questions there.

3. I don't know if you will need to know about logarithms in life, but you will need to be able to solve complex and new problems, and logarithms are an important part of learning that process right now.

4. Haven't you ever wondered what x is in the equation $2^x=30$? Well, guess what? Logarithms are just the tool for you.

5. The common core standards that were written with intentionality toward college and career readiness have stated:

   For exponential models, express as a logarithm the solution to $ab^{ct} = d$ where a, c, and d are numbers, and the base b is 2, 10, or e; evaluate the logarithm using technology.

6. When you take the standardized tests that will completely define how you think of yourself as a mathematician (despite all other messages I have shared with you all year long), there will be logarithms on it, so we better be ready!

   I could go on. After teaching for fifteen years, fielding versions of this question, and making solid online connections around such questions, I've got a whole collection of responses in my back pocket. Nonetheless, this logarithm one really hit me harder than usual. The logarithm notation is pretty confusing. The applications for logarithms exist, but I cannot say for certain that more than 10 percent of my students will actually use any of the logarithm rules in their lives. "Logarithms *are* kind of dumb," I thought about saying, joining my student in their criticality. Hey, if you can't beat 'em, join 'em, right?

Then, as I turned over the next exit ticket, another response greeted me: "I think these log rules are fun. It's like a puzzle. Can we do more of this?"

"A puzzle!" I thought. "Can we do more of this!" I guess I couldn't join the anti-logarithm team just yet, as I would be marginalizing this student's experience.

As I thought about these two exit ticket responses, I got to thinking about students' feelings about what is and isn't necessary. My guiding question became: "What type of classroom community helps as many students as possible feel as often as possible like what they did during a class period was necessary for their lives on any given day?"

"Dear Math, Maybe I won't try to figure you out; maybe I'll try to find a way around you."

IS MATH UNNECESSARY?

Unnecessary, useless, unneeded, unhelpful, pointless. These words permeate student letters to math. From explicit usage of the phrase "I don't need you" to implications in phrases like "Your answers are available online" or "I'll just find a way around you," students are often skeptical about the "need" for what they are learning. What is it that brings mathematics this level of critique? Do disciplines like history and science garner such intense criticism?

To dive into the question of why we need math, it's important that we unpack what it truly means to "need" *anything*. At its core, a need means it is necessary for survival. Most basically, individual need implies that without this thing, a person would no longer survive. Maslow famously created a hierarchy of needs, defining physiological needs at the base, all the psychological needs in between, and self-actualization at the top as the way a person might achieve their full potential. While it may be tough to justify math's physiological necessity, it's worth considering the

51

other ways we need math. In this section, we'll walk through different types of needs that my student advisory panel helped outline in our conversations.

"Dear Math, Your basics are more useful than any of your advanced stuff. To be honest, I don't feel like I would find any of your advanced things useful unless I become a scientist."

UTILITARIAN NEED

Students frequently express the desire for a utilitarian need for the mathematics they are learning. They want to say, "I will use this exact skill at some point in my future." When math is being taught in a straightforward way ("Here is a math concept; here is how you do it; now, you go practice it"), the math content is the part where students may find themselves seeking this type of utility.

While this need appears to be genuine, the realization of it can be misguided. It is impossible to predict exactly which pieces of content knowledge a student might encounter later in school, let alone in their life outside of school, in a job they might get, or in a family situation they might see. It's impossible to design curricula that could prepare all students for only their individual futures. Yet the notion that schools should provide such exact teaching cannot be overlooked because there are, in fact, many jobs that necessitate the concepts addressed in middle and high school mathematics. Exposure to these ideas may be just the inspiration a student needs to follow a career path they may not have otherwise explored. We can't ignore this type of need, but neither can we stop there.

"Dear Math, I honestly don't know how I am gonna use you in my future career because I wanna become a videogame designer, but I guess I would use you by measuring the height and width of the models I will be making."

EDUCATIONALLY IMPOSED NEED

Another need that feels quite genuine to many students is the educational need to graduate. In 1894, the Committee of Ten, a group of predictably homogenized advisors, declared that mathematics be compulsory and yearly for students. The American public school systems have carried on that guidance to this day. The colleges are on board with it, and some university systems only look at applications from students who have taken three years of mathematics. Most high schools require students to pass at least two years of math classes to graduate. This educationally imposed need is also quite utilitarian.

"If you don't do this, you won't pass, and you won't graduate."

In student letters, this sounds like "I guess I'll just do what I have to do to get by" or "Maybe I'll find a way around you."

Learning because you must in order to "get by" may have its merits (I'm hearing my father in my head stating, "Sometimes, life deals you circumstances you wouldn't have picked, and you have to figure out a way to persevere through them for the sake of something better"). A commonly cited reason for doing math is to develop this "doing something I don't want to" muscle. This, too, feels like a sad version of "need" as motivation in our classrooms.

"Dear Math, I've never liked you and probably never will. You sometimes teach me random stuff I will never use. You always pressure us in different ways when we're comfortable with one. I've had math since kindergarten, and every year there is more and more pressure. If I did have to say my favorite topic, it would be fractions."
—David, seventh grade

AMERICA NEEDS YOU TO BE GOOD AT MATH: SOCIETAL NEED

During the Cold War, America's leaders decided that pouring more money into STEM classes would help advance our country, maintain our power, and give us the tools we needed to win the war. Thus, students were presented with a societal need to learn math. As a result, people wanting to be contributing members of society had to fall in line. But is the basis for the decision true? Does a person learning math help make the country a better place? Maybe, but even this type of need has problematic elements.

The military-industrial complex relies on mathematics knowledge to design and engineer the newest, fastest, and most precise technology. This leads to discussion questions such as: What is this technology used for? Is it always for the good of humanity? As decided by whom?

Relatedly, data science has been gaining traction as an increasingly useful field of mathematics. The amount of data that exists in the world is astronomical and growing each minute. To process this data in a meaningful way, we need strong mathematicians who not only have the tools to tackle such work but also have the creativity to think of new ways to use these numbers. More discussion questions include: What types of uses are we looking for? Are they always in service of improving our lives and our communities?

Exams like the PISA are frequently cited, telling a story of the United States as "behind" in mathematics. Here's another discussion question: Can we learn mathematics to help our country perform better on such comparisons?

"Dear Math, I still see no purpose for anything that isn't the basic addition, subtraction, division, multiplication, etc. Unless you want to have a future in mathematics, then it shouldn't be taught for students who don't want that in their career."

THE ETHICS OF IT ALL

Is the lack of a competitive edge in the military and on international exams enough to create a need? In what ways is mathematics truly neutral and objective? In his book *Reading and Writing the World with Mathematics,* Eric Gutstein argues that it is not neutral at all and that ethics play an integral role in the world of mathematics and should be infused into our K–12 teaching daily.

From utility to education to society, a variety of areas exist where we might cultivate a need for mathematics. In all these needs, however, there is a clear external pressure to learn math, whether it's from a job field, a college, or the country. I find these bases for necessity to be incredibly unsatisfying and even outdated, given that the majority of jobs of the next century haven't been created yet and that computers and automation are increasingly able to tackle jobs that humans once did. In light of this, it is likely that we must reimagine a new tactic for motivation. In this next section, we will think through a different set of needs to help us operate from new assumptions.

RETHINK NEED WITH THE NECESSITATING PRINCIPLE

Beyond the utilitarian, educational, and societal needs is a set of deeper needs. One of those needs is to make sense of the world around us. While math class ought to be a place of reasoning and sense-making, the structure of classes often doesn't lend itself to that. This lack of integration with students' perspectives undergirds the feeling of needlessness. How frequently do students dive into solving a problem without knowing what the problem is about or what a potential solution might be? Guershon Harel, a math professor and researcher in San Diego, wrote that "Any piece of knowledge humans know is an outcome of their resolution of a problematic situation." He goes on to explain, "For students to learn what we intend to teach them, they must have a need, where 'need' is meant intellectual need, not a social or economic need."

By intellectual need, Harel offers a deeper understanding of the dilemma of students getting answers to problems when they don't even understand what is being asked. He goes on to state that true intellectual needs come from a problem that students "understand and appreciate."

This notion of understanding and appreciating might be linked to a utilitarian view of mathematics, such as, "I'm going to become an architect, so I need to learn about complementary angles." More often, it is connected to a human desire to make sense of ideas and continue forming new connections about things we know and think we know. As teachers, we design daily lessons and units that create this need.

THE HEADACHE

Launching a lesson with a phrase like "Today we are learning about equivalent ratios," is not an uncommon practice. Curricula and teacher prep programs tout having a clear mathematical understanding of goals and making sure students are connecting their learning each day to the stated goals. Beginning a lesson with this type of clarity can have a place and purpose, but it may not be the piece that creates a true "intellectual need." This structure relies on the teacher's authority or state standards and the students' buy-in to this authority for what they should learn to guide the lesson launch. But is this enough to create a feeling of "need"?

Certainly it is for some students—those who show up at school every day, anticipating that they will be told new information to add to their mental file folder of information, regularly preparing them for whatever is next. While this type of student might be celebrated as a "good student," I propose that mindlessly buying into learning because a teacher said you should learn it can damage a student's sense of agency, ability to construct new knowledge, and sense of criticality toward new information. Students who ask, "Why do I need to learn this?" are being helpful and critical and exhibiting the type of advocacy and agency that we hope for students to practice. The dilemma then lies in the ways that we plan lessons to attend to this skill every day. We can plan lessons in ways that encourage students and parents to ask us this question so that we can share our clear justification of what we are doing each day.

Dan Meyer, a proponent of students doing "real work" in math class, discusses this dilemma through the lens of headaches and aspirin (2015). He shares the following two questions as guides for lesson planning. Ask yourself:

1. "If this skill is aspirin, then what is the headache, and how do I create it?"

2. "How would a mathematician's life be worse if they didn't have this skill?"

In his blog post on this topic, he explains, "Math shouldn't feel pointless. Math isn't pointless. It may not have a point in job [y] or [z], but math has a point in math. We invented new math to resolve the limitations of old math. My challenge to all of us here is, before you offer students the new, more powerful math, put them in a place to experience the limitations of the older, less powerful math."

THE LAUNCH—NOTICING AND WONDERING

Necessitating the learning by creating intellectual "headaches," or reasons to solve, is an imperative in any classroom, and I applaud Meyer and Harel for guiding us with this language. I want to deepen our thinking by sharing connected ideas for creating a "need" among middle and high school students.

Often, I begin a lesson with an image on the screen. Whether it's my kids' most recent artwork, a succulent I found on my run that morning, a screenshot of my child running down the street, a puzzle, or a visual pattern, the image is meant to provoke student thinking, sharing, and questioning. I like bringing these in from my life so that I can ramp up my storytelling about the image, but loads of brilliant math educators share images like these all the time on the internet ("would you rather," visual patterns, and Talking Math with Your Kids, to name a few). Math scholar Dr. Julia Aguirre uses images to invoke thinking about a social-justice-oriented modeling task. For example, for her Flint water task, she launches with an image of thousands of water bottles on pallets in the town of Flint, Michigan. For my food desert task, I launched with an image of a map with labeled grocery stores.

Pictures can be more valuable than words in this type of lesson launch because the students can share what they are noticing in their own words and bring their experiences and knowledge to the situation. Based on Harel's description of knowledge-building, this gives them the option to add new information to their previous understanding of an idea. By asking, "What do you notice?" or "What do you wonder?" students can share their thoughts about the picture and then ask questions. As the teacher, you can select the questions they ask to guide the rest of the lesson in the direction of your mathematical goal for the day.

The Notice and Wonder routine was formalized by the Math Forum in the early 2000s, and founding member Annie Fetter continues to encourage its use through presentations and her Always Noticing and Wondering blog site.

STORIES WITH RESTATES

Who doesn't love a good storytime? Even as adults, we love stories, particularly ones with intrigue and those we can connect to. My routines for this type of launch are drawn from my work with elementary teachers who use the CGI (cognitively guided instruction) model in their schools. The model relies on problem-solving through story problems (commonly known in math language as "word problems"). In a CGI lesson, considerable time is spent at the beginning to develop a shared understanding of the story for the day. Students work together on just one problem each day. The story problem is frequently posted on a big poster at the front of the room, and one student reads the problem aloud. The students then turn to one another and restate the problem. The focus isn't on answering the question but on merely making sense of what the problem is asking. Then they read the problem out loud again and predict a potential answer or what it would look like.

The students take time to reason about the quantities in the problem and whether an answer like four or three million might make more sense. They also consider what type of units might make sense for this type of problem. Only after this breakdown of the problem do the students begin to engage in the "grapple" time on the problem. This type of storytelling can be easily modified for middle and high school contexts. Sometimes, I will post a story or a claim on a slide on my screen and ask a student to read it out loud, and then they do the same turn and talk. Helpful questions include:

- What is this problem about?

- What are we trying to solve for in this problem?

- What might a potential answer to this problem look like?

Another twist on this idea comes from Chris Weber's mathematical imagining, which advocates for a storytelling opening where the students sit back and (optionally) close their eyes while a teacher describes a mathematical scenario. After the storytelling ends, the students open their eyes and spend time journaling before beginning to share out with their group.

PUZZLES

Have you ever seen those ring puzzles on the table at a friend's house or a doctor's office? Did you *need* to solve the puzzle? No. But did you want to? Yes! We love reasoning our way through puzzling situations. And mathematics has some pretty puzzling situations. One of my favorite puzzles is WODB (Which One Doesn't Belong) from Christopher Danielson. In these types of puzzles, the students need to "construct a viable argument" for why something doesn't belong or why they would rather do one thing over another.

Another favorite twist on this is bringing in student thinking from the day before. For example, "Yesterday, one of your peers conjectured that when you multiply the top and bottom of a ratio by the same number, you get an equivalent ratio" (show an image of student work on the screen). "Do you agree or disagree? Explain." Whether or not the topic conjures utilitarian needs, the human desire to reason and make a claim of agreement or disagreement comes into play, and often, a debate ensues. Who doesn't love a hearty debate? And what could be more like "real life" than discussing your opinion on a situation, carefully listening to someone else's opinion about the same topic, and working through the evidence to figure out which is more correct and whether a possibility exists that both opinions are correct.

UNSOLVABLE EQUATION (SO FAR, AT LEAST)

Given your awareness of your students' skills and your understanding of the content progression, it can be helpful to pose an equation that requires students to use a new tool to solve. Harel claims this is when "students might encounter a situation that is incompatible with or presents a problem that is unsolvable by their existing knowledge. Such an encounter is intrinsic to the learners, for it stimulates a desire within them to search for a resolution or a solution whereby they might construct new knowledge. There is no guarantee that the learners construct the knowledge sought or any knowledge at all, but whatever knowledge they construct is meaningful to them since it is integrated within their existing cognitive schemas as a product of effort."

An example of this is at the beginning of the logarithm unit I mentioned at the outset of this chapter. I placed on the screen:

$$2^x = 17$$

What do you know, think you know, or wonder about this equation?

The students didn't yet have logarithms as a solving tool for this problem, but they were able to make sense of the structure and use what they knew to guess what the answer might be.

Our brains love patterns. Launches that involve making sense of patterns on tables or even visual patterns and asking, "What do you notice, what do you wonder, and what comes next?" give students the opportunity to make use of structure and then construct an argument and defend their solution. See Image 3.1 for an example.

3^{-1}	
3^0	
3^1	
3^2	
3^3	
3^4	

Image 3.1

It's critical to launch a lesson in a way that makes students feel like their thinking and reasoning are a needed part of the day. Whether or not the utilitarian, educational, or societal needs are in place, the need for students to think is present each day. Zaretta Hammond, educator and author, speaks of this practice as culturally responsive teaching. She states, "The power of culturally responsive teaching to build underserved students' intellective capacity rests in its focus on information processing. Processing is the act of taking in information with the intent to understand it, relate it to what you already know, and store it in a way so that you can easily retrieve it."

When we launch a problem in a way that necessitates and seeks student reasoning, sense-making, and questioning, we show students they are needed every day in every class period. Hammond goes on to state that in a lesson, "The goal isn't simple engagement but engagement so that the brain pays attention, recognizes what's coming, and lets new content be offered."

At the end of the day, the math content we are exploring might not be necessary for a utilitarian or social need (this doesn't opt us out of teaching to fill those needs as frequently as possible), but we still need to make our classes feel necessary for students each day. For example, we can convey that the class is necessary because in it, you learn more about yourself. We explore who you are as a collaborator and problem-solver. You find out how you approach challenging problems and receive feedback from others. You notice how you respond when you make a mistake. You learn how to listen well to others and connect to their ideas. You might even gain a tool to solve climate change or lessen the impacts of structural racism. Regardless, I hope that together, we can design learning that feels necessary.

WHAT'S "NECESSARY" IN THE CLASSROOM?

It was the spring of 2021, and I was teaching seniors during the pandemic. Times were hard for teachers, who were used to being student-centered and caring about the people we were tasked with supporting. Times were also hard for the students, who had been robbed of the senior year they imagined. The course was entitled Math 4, and we did conceptual work around calculus while also working on a project about the election and another project to unpack the stories we could learn by taking a mathematical look at our neighborhoods. I determined this was what the students might "need" in this particular semester. We had our weekly exit tickets to check in on learning and feelings about math and about the pandemic. I wrestled through feedback from students that semester, even more so than usual, as it was hard to meet all students' needs. The students shared what they needed "more" of:

- more calc work
- more work time in class
- more rigorous math
- more "real" math
- more lectures with the paper under the doc cam
- more Flipgrids
- more hands-on things

Then the students shared what was working for them:

- the schedule
- the workload
- the content
- working with partners
- working in groups
- learning about topics outside of school and relating them back to math
- always getting help if I need it

On the final question labeled "Anything else?" I read a response that made me feel like all was right with the world, if only for a moment. The student feedback stated:

"This is the only class I feel like a human being in."

Even though I can't do exactly what every student needs every day on every dimension of "need," I could foster a sense of our collective humanity even in the direst of circumstances—and what could be more necessary than that? This sort of connection with students is integral to teaching. Parker Palmer, founder of the Center for Courage and Renewal, observes, "Education at its best—this profound human transaction called teaching and learning—is not just about getting information or getting a job. Education is about healing and wholeness. It is about empowerment, liberation, transcendence, and renewing the vitality of life. It is about finding and claiming ourselves and our place in the world." We are at our best when we empower our students to find their purposes for learning math.

GIGI'S REFLECTION

Despite the obvious evidence to the contrary, math is not my passion (I know I just became a whole lot less credible as a math education author, but stick with me for a second). My most vocational passion is writing, a skill that is not inherently mathematical. Because of this, my scholastic career has consisted of a couple of seemingly unnecessary subjects: biology, PE, and—everyone's favorite to chastise—math.

At the root of everyone's mathematical journey, there is a clear necessity. Addition, subtraction, multiplication, and division are accepted as universally beneficial. Linear equations and geometry also have a pretty strong "necessary" consensus. But by the time we make it to calculus, most people, other than those working as SAT tutors, have a tougher time explaining the necessity. In some ways, they're right. I have always had a certain appreciation for mathematics, yes, but without aspirations of becoming a statistician, seismologist, or accountant, many of these topics have had the initial impression of unnecessary. It is true that I'll most likely never have to calculate an integral or solve a quadratic equation. When it comes to pure life preparation, taking these topics at their absolute values does not yield necessity. Where learning them does become imperative is within their secondary skills.

For instance, in my exceedingly strange and, if I'm being honest, lackadaisical senior year, we tackled myriad math projects. We'll dissect two for this analysis. In one, we investigated the math behind different voting systems in place around the world and throughout time. This project was necessary for easy-to-identify reasons. It was powerful, poignant, interesting, and, to call back to the words of another of Sarah's students, it made me feel like a human. The second project was calculus-related. As amazing a teacher as Sarah is, when I was staring at the curve of a derivative on my computer screen with dark circles under my eyes, a blanket wrapped around my body, and a pandemic raging on outside, calculus did not feel very interesting, and it certainly did not feel necessary. I found myself absently following along through lessons and lazily working through assignments. This was until I had the revelation of discovering the secondary skills of calculus. The primary skills are obvious: limits, derivatives, and integrals. These are all pointless for most people (sorry if that's harsh to any calculus teachers out there). The secondary skills, though, lie not within the content but in the learning process as a whole.

Collaboration, self-motivation, and time management are unequivocally useful skills that everyone, regardless of prospective career paths, can absolutely benefit from refining.

That semester may have been a slightly lazy one for me, and I think I would have serious trouble defining a limit at this time, but I got better at asking my peers for help, and I had to practice my self-management. These skills are not specific to calculus, and they apply to any subject that challenges us to persevere with the learning despite a seemingly invisible need for it. I am certain that I will continue to refine and appreciate these skills as I move through my life. I think they may be even more necessary than addition.

INVITATION TO
REFLECT

- What made you feel like you "needed" to learn mathematics when you were in school?

- In what ways have you seen your students experience a "genuine" need to solve a problem? What was happening in your class that made it feel truly needed?

CHAPTER 4

DEAR MATH,
YOU ARE INTIMIDATING

*"Dear Math, When I stood up and I couldn't
explain a problem, I was so embarrassed."*

TAMARA WALKED INTO my class on the first day of her sophomore year. A transfer from across the country, she was
new to both our school and the area. While she had left
a lot behind when she moved from the East Coast to the West
Coast, one quality she brought with her was an intimidation when
faced with math. This feeling of timidity became evident in the
first few days. While generally precocious and independent, when
it came to exploring mathematical ideas, Tamara rarely shared
her thinking in her groups. She would kindly listen to others, ask
thoughtful questions, and then avoid raising her hand when class
discussions ensued.

She shared her concerns one day after class when I approached
her about how important her thinking was to our class.

My usual response of "I understand … and I hope you come to
see that this class is a better space when mathematicians like you
are participating wholeheartedly" was taken in, but she responded

with skepticism. Her Dear Math letter from later that week told a deeper story:

> "Dear Math, Oh, do I have some things to say to you. You've followed me throughout every school year, caused me the worst headaches, and given me numerous counts of anxiety just thinking about you. The memories of my seventh grade math teacher telling me, "Maybe you're just not a math person," still ring in my head, and the constant B's and C's are imprinted in my mind. You've been a never-ending challenge and struggle, and it's always been hard to understand you. No matter how many times my friends and teachers explain you, I never grasp you completely." —Tamara

I knew that from this moment forward, my work was to get to know the mathematical strengths of Tamara and share them boldly with her peers. That needed to start by dismantling the beliefs that Tamara had internalized about herself many years before: that she was not good at math and was not a math person. I also knew that she needed to find mathematical anchors in our daily classroom work that weren't grades on tests or final grades for classes. The feeling of intimidation was fierce in this student. Even though she was confident and curious in her life outside of math class, she hadn't yet found a way to bring those qualities to class with her.

WHY THE INTIMIDATION?

If something is intimidating, it makes a person feel fearful or timid. In student letters, we sometimes see the explicit word "intimidation," but it often comes across in words like frustrated, inadequate, nervous, embarrassed, anxiety, scary, and annoying. Students speak of various activities that brought on feelings of intimidation, like being asked a question they didn't know the answer to, the length of assignments, and getting low grades on tests. Another interesting theme that comes up in letters is a teacher's message of "preparation for something that is going to get harder." This language gives students the feeling that difficulties are on the horizon, and they need to get ready so they don't fail.

> "Dear Math, I was in kindergarten, sitting around my teacher with all of my classmates. She asked a simple subtraction question, and I confidently raised my hand to answer. I knew I was right. I had to be. I was wrong, and it was then that I knew shame. My face turned red, and the lump in my throat was threatening to break. I couldn't tell if I was more ashamed that I was wrong or that I was about to cry. Since then, I've had a paralyzing fear of being wrong—especially when it comes to you. I've always tried extra hard to make sure I'm a step ahead of you, and when I don't get one of your concepts immediately, I panic." —Tenth grader

While students address math directly and speak to the ways they are intimidated by it, the underlying stories they tell speak not of math as a discipline but of classroom experiences around math and the ways that mathematicians share in problem-solving that caused this feeling of intimidation. When I first started having students write Dear Math letters, I anticipated hearing stories of the intimidation of the symbols and "confusing math speak," but I was surprised to hear that most of their intimidation origins had more to do with classroom practice, home experiences, and societal pressures rather than with the actual subject of math. These themes are congruent with research on math anxiety, which suggests that it stems from these same three spaces: home, society, and classrooms (Whyte and Anthony 2012).

Home

The stories that students are told about math at home can come from parents being doubtful of their own math abilities or their own feelings of intimidation by the subject; they say things like "I was never a math person" or "That math is so far above my head, I can't even help them anymore." This normalizes the feelings of confusion in a way that isn't productive (for insights about productive confusion, see Chapter 11). Other people hear stories at home about how their parents were good at math "back in their day" and how math these days just isn't what it used to be. These messages, too, can undermine the student's work and cause them to feel intimidated by what their teacher is asking them to do because it implies that maybe they aren't doing it "right."

Society

The societal messages about math intimidation are real as well. Countless messages come across our TV screens each day where people make jokes about failing math tests, talk about how math is confusing, and reference math as a thing that "nerds" do. These messages can add to the intimidation factor because they make it seem like this is "the way" to be smart and that if students don't prove themselves in math class, then their smartness might be fake or fraudulent.

Classrooms

Our classrooms can be the places that bring about intimidation. As students mention, the need to perform on tests quickly and with correct answers serves to intimidate students. Even worse, some students speak to feeling like math was presented in a way that only a select few will be able to access. Though often used jokingly, the term "proof by intimidation" may be a frequently used tactic in classrooms. If a teacher is explaining a challenging math idea to students and glossing over the details, waving their hands around as if the ideas are obvious, and then moving on, they're "proving" that it works and is simple, and everyone should just understand. If a student is sitting in the classroom and doesn't understand, they can feel intimidated. This "proof by intimidation" happens at the expense of students being able to do any sense-making or reasoning. This type of teacher behavior is common and may be an underlying factor for the ways that math feels intimidating to students.

STOKING THE FIRE OF INTIMIDATION

In their letters, students share language like "I was so scared I wouldn't know how to do a problem" or "When I had to share my thinking with the class, I panicked," which indicates a belief that one either "has math knowledge" or "does not have math knowledge." The notion of not having math knowledge is quite intimidating, even though it is perfectly normal to not have all the knowledge about something. Again, the theme of a "fixed mindset" is cropping up here. I recently heard a story about a young student who, by all standard measures, was a successful math student. She got a five on her calculus AP test and got all A's in every high school math class, but when she got to college, she was so intimidated by her first math class that she vowed to never take another one. Unfortunately, this story is a common one.

"Dear Math, I want to feel that passion of loving you again, Math, but that barrier of the word problems is stopping me from doing so. I basically feel as if I'm stuck behind a locked door because of that. It's why I tend to give up on myself and not participate in class because of the challenge to look smart and know what I'm learning and not look stupid." –Dylan

HOW INTIMIDATION LEADS TO LOW PERFORMANCE

As teachers, we are necessarily holding power in the classroom. We hold power to assign activities, call on students, reprimand students, give grades, and decide what's on a test. We get to decide how to use this power.

In society, math holds power. Math is often upheld as the ultimate way of knowing something. It is seen as the path to intelligence. It is used to keep students out of some classes and usher them into others.

The combined power of the teacher in a classroom and math in society can be a recipe for disaster (deep intimidation, poor relationships, traumatized identities, and low achievement) or be leveraged for good (communal thinking, shared sense-making, and beauty-seeking).

> "Dear Math, When I was younger, I was very good at math, and in fourth grade, I won an award for math. And in fifth grade, my teacher was very very hard on the students, and she would tell us that sixth grade will be harder. And it's hard, yes, but it's a little easy." – Seventh grader

When teachers cause and sustain math intimidation, it makes the entire field of mathematics worse off. When we create environments where students are intimidated, we are getting far from the best from them. As Sarah Sparks wrote in her *Education Week* article, "Math anxiety is linked to higher activity in areas of the brain that relate to fear of failure *before* a math task, not during it. It takes up mental bandwidth, causing discomfort. This discomfort leads to avoidance and ultimately results in a worse performance on any type of math activity." If students are entering our classrooms with feelings of intimidation before they even begin, it is important that we intentionally design our classrooms to disrupt this intimidation from the outset of each class.

> "Dear Math, I know that with practice, I could get better. But currently, I am practicing and trying my best to get better, but it isn't working out so well. In the future, I hope that my mental math gets better, I

can understand math quicker, I can do math quicker. I also want to be able to not get so nervous when a test comes up and end up having a bad grade." —Seventh grader

The first step is to model good use of power by the teacher. This gives the students an anchor by which they might wield their power. The second is explicit celebrations of lots of ways of being mathematical in our classrooms so students see that there isn't one main holder of the information in the classroom and that they needn't be intimidated by their peers. In addition, structures like the Daily Discourse, which place power in the hands of the students, create a culture that gives space for students to truly listen to and value one another's contributions.

DAILY DISCOURSE ROUTINES THAT OVERCOME INTIMIDATION

Because intimidation implies there is a problematic power dynamic between the participants—in this case, the teacher, the students, and the mathematics—attempts at dismantling that wielding of power must be intentional and frequent. In this section, I'll share a routine I have been iterating with my students over the past fifteen years. It has taken on several different names over time; however, the most recent iteration was named Daily Discourse, as it spoke to the frequency of the routine and the central goal of the routine, which was a discourse about mathematics. The goal of the discourse is to situate mathematics as an idea-generating, collaborative, exploratory, nonpunitive, and nonperformative activity. It also shares potentially confusing math equations and symbols as ponderable, explorable, creatable, and discussable, not as something to be known or not known.

Admittedly, Daily Discourse is not a failproof solution to overcome the misuse of power that leads to intimidation. Even if the teacher successfully excavates and amends their use of power, classroom dynamics can play out, and peers can serve as the primary source of intimidation. This takes work to resolve, but, as discussed in Chapter 2, it is entirely possible.

In a blog post in early 2021, Geoff Krall wrote about "the next iteration in classroom conversation orchestration." He shared the deep value of frameworks like "the five practices," which center student discourse in a teacher-facilitated way. He claimed that the next iteration in the process toward becoming more student-centered would be the teacher stepping back to allow students to facilitate their own discourse. He asked, "How could we foster even *more* student ownership of the conversation? It almost never happens naturally. Students typically have to learn how to drive the conversation, or at least move it forward." He goes on to share the work that elementary education authors Kassia Omohundro Wedekind and Christy Hermann Thompson have done in this area by working with young students. Their work is provocative, and as is often the case, we at the secondary level have much to learn from our elementary colleagues. Similarly, this type of work has been heavily researched in the humanities world, where Socratic seminars—student-led, student-centered dialogues—are gaining traction. As I started designing my new warm-up routine, I knew I wanted it to be wholly student-centered. While maintaining that focus, my Daily Discourse design has evolved through the years.

WHAT IS DAILY DISCOURSE?

Daily Discourse is a ten-to-fifteen-minute activity that encourages students to engage with abstract math problems in a way that facilitates collective sense-making and builds from what the students know and think they know. Multiple solutions, ideas, and questions are generated through student-facilitated discussions, emphasizing that everyone has something to contribute, that there are different ways of being "mathematically intelligent," and that all are needed to progress our collective ideas.

THE DISCUSSION LEADER

As Krall mentioned, having students lead the discussion does not come naturally and will take intentionality. For the first few weeks of the school year, I usually start by leading the discussion myself. After the discussion, I ask questions like "How did we do in that discussion? What went well? What could we improve on?" They speak to their actions, but then I ask follow-up questions about my actions that supported a strong discussion. After a few weeks of scaffolding in this way, we start the routine of having student discussion leaders. I first take volunteers and allow students to opt in to lead the discussion. If the student is comfortable, we have a meta-moment afterward to discuss what the person did well in their leading and to celebrate their contribution.

After a few more weeks of this, I announce that all students will be taking the opportunity to lead discussions throughout the remainder of the semester. We co-create a rubric around expectations for a discussion leader (see Image 4.1), and students get the rubric and keep it in their binders. After they lead the discussion for a given day, they do a self-assessment of their work as the discussion leader.

SAMPLE CO-CREATED RUBRIC
FOR DISCUSSION LEADERS

Circle which of these you achieved today	Put a checkmark next to actions you were able to complete today
EE (exceeds expectations)	☐ Asked clarifying questions ☐ Got EVERYONE involved (whatever it takes!) ☐ Encouraged building off of ideas ☐ Encouraged people to share who we haven't heard from in awhile ☐ Asked people to expand on their thoughts ☐ When a question was asked, they asked if anyone had an idea about it ☐ Kept it smooth and flowing well (stayed on topic) ☐ Was loud and clear
ME (meets expectations)	☐ Got a few people involved ☐ Some conversation was smooth, but other parts were awkward ☐ Saw very little building of conversation, just random ideas ☐ Was a little hard to hear ☐ Asked 1–2 follow-up questions
NY (not yet)	☐ Stood in the front of the classroom ☐ Asked people to share an idea ☐ Conversation in the room was awkward/silent ☐ Rude comments made toward classmates
I (incomplete)	☐ Did not act as discussion leader

Image 4.1

COMPLETELY OPTING OUT

Admittedly, the idea of standing in front of a classroom of their peers is stifling to some students. Even with intentional scaffolding, I give a way for students to completely opt out. From my experience, however, this is rare if students receive sufficient encouragement, clear expectations, and scaffolding. Nevertheless, if a student is persistent that this isn't for them, I offer the "scribe" job (described in the next section) or an alternate assignment to cover the appropriate bases; for example, they could offer written feedback to a discussion leader on a given day or volunteer to guide the discussion at their individual table.

More often than not, however, students rise to the challenge, and some students completely thrive as a discussion leader. While the charismatic students generally do well, I have frequently been surprised by students who don't step up as leaders but do well in listening to student responses, asking follow-up questions, and weaving a beautiful discussion together with their peers.

THE SCRIBE

Another key component of the student-led Daily Discourses is the scribe. This person stands up at the board with the discussion leader—which is also helpful for reluctant discussion leaders—and transcribes what is shared. For many students, this role is more challenging than the discussion-leading role because it requires listening well to their peers and synthesizing what is shared into something quickly writable on the board. Still, some students thrive in this job, and I often get a few students who are not only excited to scribe but start getting nominated by their peers to scribe because of the ways they are helping the learning unfold in the classroom. With this role, I encourage all students to try it once and reflect according to another co-created rubric. Similarly, after a scribe completes their job, they do a short self-assessment on the rubric, which they keep in their binder for sharing when they submit their portfolios. See Image 4.2 for an example of a scribe scoring rubric.

SCRIBE SCORING RUBRIC

Circle which of these you achieved today	Put a checkmark next to actions that you were able to complete today
EE (exceeds expectations)	☐ Got important/main ideas down ☐ Summarized what people said ☐ Made things understandable/simple on the board ☐ Used clear notes/pictures/drawings! ☐ Asked clarifying questions when you don't know what the person is saying ☐ Asked the person to draw their idea if it is complex (invite them up) ☐ Kept information organized on the board ☐ Readable handwriting (as best as possible)
ME (meets expectations)	☐ Wrote some ideas/skipped a few ideas if they were confusing ☐ Wrote a few notes and a few drawings ☐ Information was fairly organized on the board ☐ Handwriting was okay, perhaps hard to read from far away
NY (not yet)	☐ Tried to write complete sentences ☐ Didn't ask questions when they were confused or invite people up
I (incomplete)	☐ Did not write anything on the board

Image 4.2

OVER TIME

If you are doing a similar style warm-up for several days in a row, have a few "extension prompts" in the back of your head for students who are ready to take their brainstorms further. You can also brainstorm ideas every few weeks on what students can do if they think they have "finished a problem." See Image 4.3 for student-generated ideas.

Have students track and reflect on how long they spend brainstorming each day, when and why they were at their best, and what they can do to engage with the problem for at least five minutes. Then, identify examples of successful brainstorms from student work. Periodically share them with the class to help them unpack what a successful brainstorm looks like and what successful brainstormers do.

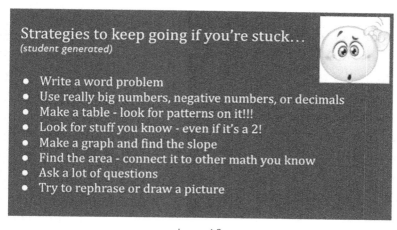

Strategies to keep going if you're stuck...
(student generated)

- Write a word problem
- Use really big numbers, negative numbers, or decimals
- Make a table - look for patterns on it!!!
- Look for stuff you know - even if it's a 2!
- Make a graph and find the slope
- Find the area - connect it to other math you know
- Ask a lot of questions
- Try to rephrase or draw a picture

Image 4.3

THE TEACHER'S ROLE

The teacher, still holding the given power of the room, is now using their power to intentionally avoid intimidation and create a community of learners. This doesn't mean, however, that the teacher has nothing to do. Throughout the years, I have experimented with various uses of my time. Here are four iterations.

Iteration 1: First, I placed a check mark next to the name of every student who spoke, and I shared this data with the students, saying, "We are all needed in this community. How did we do today in terms of hearing all the voices?"

Iteration 2: When students started sharing for the sake of sharing (like to simply say they noticed there was a two in the equation), the class and I decided to up the annotations of my sheet to reflect the type of sharing students did. They helped me decide how to annotate the different ways they share observations (check mark), ask questions (question mark), and connect concepts (the letter c). Again, I would share this data with the students with a reflection question about how they were doing individually and how the class was doing all together.

Iteration 3: For my current iteration of this routine, I decided to hand even more authority to the students. Mimicking a practice I learned from my humanities colleagues who use Socratic seminars, I started a routine where each day, half of the class was sharing, and the other half was observing. The observers had forms on which they documented ways that their peers participated and helped further the mathematical thinking in the classroom. They celebrated the sharers and offered suggestions on qualities to improve, like body language, questions they could ask, and moments when they did or didn't write on their paper.

Iteration 4: A colleague of mine, Joe Bandini, built out this routine with his students in a way that allowed students the opportunity to be even more intentional in their discussion leadership. During the table share-out time of the routine, the discussion leader would roam around the room, listening in at each table. They carried a clipboard and would take note of the students who were sharing interesting, unique, or insightful ideas with their table. After the five minutes were up, the discussion leader would head to the front of the room and start by calling on a few folks they had heard from as they circulated the room. When I observed, a student said something like "Sydney, would you mind kicking us off by sharing the way that you drew a picture for this problem?" With this small move, the student discussion leaders were tackling the initial "five practices for orchestrating student discussions" work and selecting and sequencing their peers' thinking. Just brilliant, Joe!

For more details on Daily Discourse and ways that students might participate, see Appendix C.

OVERCOMING INTIMIDATION IN THE CLASSROOM

Vlad was a unique mathematician. His peers knew him as one who would study math textbooks for fun, stay after class to engage his teachers in discussions of math topics far outside of what was being studied, and take college math classes in his spare time. His mathematical brain was always churning. A student like Vlad could easily intimidate his peers. A classmate with such a deep understanding of math might seemingly be ready to squash you with their math knowledge. Fortunately, Vlad didn't use his math knowledge in that way.

One day during our Daily Discourse, it was Vlad's turn to be our discussion leader. He had been late to class, so he was seemingly unprepared for the task.

"It's okay; just give me a second to grab my notebook," he said. "I was working on something last night that yesterday's Daily Discourse had got me thinking about."

Vlad came to the front of the room to share his thinking with his peers. He shared how a problem we had been working on the day before in class had made a question pop into his head, and he couldn't stop thinking about it until he put pen to paper and did some exploring. He casually shared both his mathematical thought process and the reasons behind his exploration.

Knowing that many of his peers can get intimidated by his sharing, Vlad was careful to keep the big picture of his query in mind while skipping over some of the details that might lose his audience. He said, "Then I did a bunch of calculations that don't matter too much, but the point of the story is that I was able to show that the surface area of the taurus is the derivative of the volume." He went on to say, "If you're more interested in the work, I can talk about it after class, but my main hope is that you see why it comes in handy to think about math in your free time … because you're ready to be a discussion leader all the time!"

Vlad's casual way of talking, his ease in sharing his first-draft thinking, and his process helped normalize for the class that, even for him, an apparent "math whiz," math is a journey and a process, riddled with mistakes and *lots* of first-draft brainstorming. He didn't use proof by intimidation, nor did he get caught up in all the details. He broadly shared what he did and then invited questions. May more student mathematicians share their curiosity and exploration while intentionally noticing how their sharing invites their peers into a deeper conversation rather than intimidating them!

GIGI'S REFLECTION

So much of school is intimidating. Students deal with complex essays, scary teachers, newfound independence, subconscious crushes, nonstop hormones, and worst of all for many students, math. Something is so unequivocally and quintessentially daunting about the pressures associated with mathematical correctness. For most students, the intimidation of math surpasses that of everything else on the list. I believe this is due to the combined stress from math's academic and social implications. This single subject can incite academic distress with its reputation of being the most-difficult-to-pass class, and it can simultaneously cause equal intimidation with social anxiety around being incorrect in front of one's peers. It's a perfect cocktail to achieve adolescent fear. I, like every other student I ever went to school with, fell victim to this intimidation during many lectures. Even in instances when I felt confident that I knew the correct answer, my fear of being told I was wrong in front of all my classmates eclipsed my excitement for the validation I would receive if I were right. Not only does this create a boring group discussion in which the teacher is the only one talking because no one feels comfortable enough to add anything, but it also results in a suspension of mathematical growth. It all loops back to that all-encompassing need to be "correct."

The antithesis of classroom intimidation is encouragement. Starting in my ninth grade class, Sarah implemented a practice that took place at the beginning of each class: Daily Discourse. For a moment, put yourself back in your freshman-year shoes and imagine you've just walked into math class on the first day of school. You're fidgeting with the new clothes your mom bought for you, you're trying your absolute best not to show how nervous you are, and someone in your class still hasn't gotten the memo that it's time to be wearing deodorant every day. You step foot into math class, and waiting for you on the whiteboard is a math problem you're immediately tasked with solving. Great. To add insult to injury, you're meant to work with a random group of kids at your assigned table. Perfect.

As unenticing as the practice may initially sound, it does wonders for banishing the inherent stink of intimidation that exists within the walls of any math classroom (or maybe that was just the BO from the anti-antiperspirant student). Daily Discourse begins with a silent five minutes in which every student takes time to unpack the given problem on their own. However, this step has a slightly unconventional expectation: no one is required to solve the problem. If a student feels inclined to do so, awesome! But the exercise instead focuses on students' identifying what they know and what they think they know, wondering about the problem, and hopefully eliminating the academic anxiety of being graded only on correctness.

After the five minutes are up, students then spend another five minutes discussing those certainties, not-so-certainties, and wonderings with their table group. This conversation, for me, had dual positive implications. First, I was able to hear all the other ideas that could be derived from this singular problem, pushing me to an even deeper understanding of the concept. Second, it allowed me to dip my toes into the sharing-out-loud pool. That intimidation of being wrong, although lessened by the practice of celebrating wondering and curiosities as much as answers, had lived inside of me for my entire scholastic career and couldn't be entirely ousted in one fell swoop.

Next, a discussion leader and a scribe volunteer are chosen. (Surprise! It's not the teacher who leads the discourse.) This step, small as it may seem, further encourages student engagement and participation. Finally, the whole-class discussion commences, and students have all the tools they need to feel confident about sharing. They have initial ideas of their own; supplementary thoughts and assurance from their table share-out; and collective celebration for answers, wonderings, and connections from class discussion. Wouldn't freshman-year, new-clothes-you appreciate that? Plenty of school issues cause intimidation, and math certainly doesn't need to be one of them.

INVITATION TO
REFLECT

- Who in your class has exhibited behaviors of feeling intimidated?

- What will you do tomorrow in class to excavate some of their mathematical brilliance and share it with them and others in the class?

CHAPTER 5

DEAR MATH,
YOU ARE OPPRESSIVE

*"Dear Math, When I apply for college, my GPA
is going to be really low because of you."*

I WAS RIDING ON a bus with a group of vibrant ninth graders on our way to a field trip to interview folks as part of a project they were working on about telling stories across borders. The students were generally happy and connected, talking about the past weekend and what they were looking forward to on the trip. The sound level was fairly muted until someone shouted, "Grades are posted!"

The hum of activity paused as students checked scores on their phones. Even though I was not a student nor their teacher for this particular math class, I felt a sudden panic. Just the idea that grades were posted somewhere publicly brought back a frenzy of emotions from my own experience receiving grades.

"I got sixty-five out of sixty," one girl exclaimed. "I knew I got that extra credit one right." She sat back in satisfaction for a moment before turning to her neighbor. "What did you get?"

Her neighbor smiled and said, "I got sixty-one, not as good as you, but I'm still happy." They proceeded with the conversation they had been having before the grades were posted.

Another student at the front of the bus stood up and danced over his perfect score, and then another student who got fifty-five out of sixty and was sitting next to someone who got fifty started mocking, "Haha, I'm smarter than you. Idiot."

Some students talked quietly to their neighbors about problems they remembered, and many students sat silently.

A few students moved to different areas of the bus, tears in their eyes. I wondered what I should do. Should I go sit with these students and listen to their lament? Should I try to understand their stories of not remembering what they were supposed to in the moment, of making silly mistakes, of having a hard morning with their mom and being unable to focus on the test, of feeling lazy and not studying, or of not caring whether they succeeded or failed because, either way, they were already going to summer school?

I ended up just sitting and observing for a while. The posting of the scores changed the atmosphere on the bus from lively and social students to a group of people who had ranked and compared themselves. *Was it worth it*, I wondered? I longed to go back to twenty minutes prior when we were just a group of people enjoying a field trip and the weight of bad grades hadn't fallen heavily on their shoulders.

OPPRESSION TYPES

In the previous chapter, we looked at classroom practices that cause students to feel intimidated; here, we speak to broader systems in our math classes that cause students to express sentiments related to oppression. This chapter focuses on grading and assessment systems because that is what students usually cite as evidence of their oppression.

Oppression is defined as unjustly inflicting hardship and constraint, especially on a minority or other subordinate group. Also, oppression is a pressure that weighs heavily on the mind or spirits, causing depression or discomfort. We will operationalize this definition to specifically refer to the ways that math has caused a student to be held back from accessing something. This goes beyond the individual feelings of fear to a power that math class holds over students so that they cannot succeed either in an individual course or in the career path they had hoped for.

In the student letters, we see clear evidence of this in the following ways:

Naeem and Gianna mention an oppressive fear of failing tests, particularly high-stakes standardized tests like the SAT, and the ways that failing them will cause hardship.

Sebastien and Torrien reference an oppressive power that math holds over them, using words like "You allowed me to pass" or "You have done plenty to me." These words indicate that the students don't feel agency to participate, contribute, and bring themselves wholly to math class; rather, it is being done to them.

> "Dear Math, Thank you, Math, for allowing me to pass you and your class. Thank you for allowing me to understand somewhat of what you give me to learn. We might hate each other, but we sure do work well together." —Torrien

James speaks to a deep level of oppression, one where math has undoubtedly held him back from his hopes and dreams for the future.

Given that oppression most frequently plays out in "minority or subordinate groups," it is necessary that we introduce different identity markers of race and gender at this point to further unpack the oppressive conditions of math classrooms.

RACIAL JUSTICE IN MATH CLASS

While all students are capable of doing math at high levels and contributing meaningfully to classroom spaces, the data clearly shows that not all student outcomes reflect this. Disproportionate numbers of black and brown students fail math classes, and further down the road, scholars of color are underrepresented in colleges and academia. This causes me to pause and consider the reasons behind this. Is it possible that oppressive acts are currently taking place in math classes? Even more broadly, is it possible that math education had racist underpinnings from the beginning? Let's dig into this a little more deeply. Dan Battey and Luis Leyva of Vanderbilt University's Peabody College of Education have explored the root cause of our racial divides in mathematical success in the United States and found that these spaces are designed as white spaces and, as such, perpetuate white supremacy culture, not allowing for equitable success across racial subgroups.

According to Battey, quoted in Melinda D. Anderson's article in *The Atlantic*, "How Does Race Affect a Student's Math Education?" there are ways in which math teachers, math educators, and math researchers "are perpetuating racism in schools." He states, "Naming white institutional spaces, as well as identifying the mechanisms that oppress and privilege students, can give those who work in the field of mathematics education specific ideas of how to better combat racist structures." Danny Martin, another prolific researcher in the ways that racism plays out in math education, calls all mathematics educators "to review and critically analyze how the construct of race has been conceptualized in mathematics education research, policy, and practice." In response to oppression, Martin says, "We view liberation as a means to a radical end rather than an end in itself. We imagine a world in which our relationality is not to whiteness, anti-blackness, systemic violence, a world in which we are not defined by survival, resistance, and a fight for freedom. We imagine a world in which we define ourselves, our joys, and our desires in infinite multiplicities and in which we are committed to individual and collective black fulfillment." While slight modifications to our teaching practice might result in some change, Martin acknowledges that these changes may continue to oppress unless drastically upended and resisted.

Certainly, this type of change is both necessary and challenging, and all teachers must engage if we are to truly combat the oppression we see in the students' experiences expressed in their letters.

> "Dear Math, The reason why I say all this is because when I apply for college, my GPA is going to be really low because of you. What can I do, how can I just move on, what's next? I'm on my own in this world with nobody to count on due to me and only myself dropping the ball during these last two years of school. My main goal for high school was to head to college with a full ride because my parents will be unable to carry me on for college. I will be unable to succeed because of you. I cannot fulfill my dreams because of you. And most importantly, I cannot set my standards to continue this journey because of you." —James

GENDER JUSTICE IN MATH CLASS

While the racist constructions of math education and their implications in our current landscape are clear, other oppression can be seen when we break down the field by gender. According to the National Girls Collaborative Project, women took home 57 percent of all bachelor's degrees in all fields in 2013, yet women earned just 43 percent of the degrees in math and just 19 percent and 18 percent of the degrees in engineering and computer science, respectively. Curricula and classroom practices commonly oppress girls in their opportunities for connection, contribution, and further study in their math classes. Many of the women in my own family have experienced such oppression in mathematics courses, and they say it stopped them from pursuing STEM fields.

Grandma Richardson

My grandma shared her math story with me. She said that she liked math until algebra, where she got a D, and despite her efforts to attend after-school tutoring, she couldn't find a way to pass. She recounted that story with complete clarity, even though it had taken place over sixty years ago. She remembered her feelings in math class and the ways that the teacher who was grading her made her feel inept and not good enough. The irony of this story is that years later, she and her husband (my grandfather) started a data entry company in the Bay Area in the early days of formal data collection and analysis. She spent her entire career entering data from surveys that Stanford University administered. I wondered as she shared her story with me how her creative mind might have innovated the data collection and analysis process had she been afforded the opportunity to see herself as a mathematician back when she was fifteen.

Leanne Richardson

My sister-in-law and I chatted about her math story, which has a lot of overlap with my grandma's story. She shared that she had wanted to go into the medical field and become a pediatrician since she was a young child. She carried this dream with her until she got into high school algebra. When her algebra teacher told her that she was not good at math and would struggle to major in anything STEM-related, she changed her career path immediately. She is now a kindergarten teacher and does a beautiful job caring about our youngest learners, but I was struck by the major shift in her career path due to one oppressive message given to her by a teacher.

Karrie (my mom)

My mother had a negative math story, and I grew up hearing tales that usually centered around the basis that she "never made it out of eighth grade math." She believes that she never passed a math class in high school. When confronted with even her grandchildren's math assignments, she defers to someone else in the house. This is all despite the fact that she is a regular chef, baker, strong reasoner, and exhibits many attributes of an amazing mathematician. She didn't go to college, and I have often wondered whether her math journey played a role in that decision. Her life path is beautiful, including many years staying home with my brother and me, working for a few years in my grandma's data entry business, and working for the local parks district. Still, I wonder about the ways that mathematical empowerment and liberation would have opened more doors for her, or at the very least, given her confidence to work through math story problems with her grandchildren.

These three women in my life were the receptors of mathematical oppression, constrained in their choices by the outside force of math education. What could these women have done with more whole math identities and broader ways of seeing themselves as mathematically brilliant? Is there a way that our students, particularly those who are female or of color who have been systematically held back by math classes and have failed at extremely disproportionate rates, can be liberated from this stronghold?

While all the practices in this book center on student identity and storytelling, most of the oppressed feelings by students (and adults) are grounded in grades. The direct connection to these stories seems like the most fertile ground to begin our quest to liberate the students in our math classes.

Odds are that if you start listening to the math stories of the students in your classroom and also family members, people you sit next to on airplanes, and folks you run into on the street, you will hear broken stories frequently grounded in some grade they received at some point that helped them decide that math (and any adjacent field where they would even have to consider numbers) was out of the question for them. It pains me to think of the number of people who have shifted course because of these messages. I wholeheartedly believe that we must do better at creating classroom spaces that include and liberate rather than exclude and oppress.

Let's start by thinking about assessment and grading processes that can help.

TOWARD LIBERATION

"We cannot say that in the process of revolution someone liberates someone else, nor yet that someone liberates himself, but rather that human beings in communion liberate each other." —Paulo Freire

The work of liberation in education is extensive and multifaceted, including explicitly addressing systemic inequities, the ways that perceptions of different racial groups have formed teaching work, injustices through a mathematical lens, and the ways that problematic power dynamics play out in our classrooms and how teachers disrupt them.

While some of those themes will crop up in other chapters, this chapter begins the process of liberation by interrogating our assessment and grading practices, particularly noting how grading and assessments can be situated within the model of education that Freire proposes, wherein people in like communities are "liberating each other." I have prioritized assessment not because I believe it is the most important factor but because my students regularly cite grades and tests in their Dear Math letters, and I believe that as a white female teacher committed to the work of equity and liberation in my classroom, it is a good starting point. It must not be the ending point, as I will discuss in subsequent chapters on modeling and project-based learning, but it is a necessary element of each of those concepts.

THE IMPACT OF TRADITIONAL TESTS AND QUIZZES

Traditional tests and quizzes often make students feel like they have to do problems one way and limit their thinking to a "just do what the teacher tells you to do" mentality. Furthermore, the summarization of the performance on a test to one letter grade or number creates a divide and an opportunity to oppress creative ideas and the formation of a whole math identity.

> "Dear Math, Whenever I talk to my friends about you, we usually talk about the same things, our great fear of the SAT and concern over whether we will be ready when the time comes." —Naeem

By opening up assessments and focusing on growth and connections between ideas, we can gain insights into students' mathematical thinking … which is brilliant and beautiful! As we foster students' "mathematical voice," we can better understand how they are thinking about mathematics and what areas they might grow in next. When assessment is done in this way, it also helps guide instruction as the student ideas become the narratives that push the math learning forward. Including the ideas of all learners is not only important but necessary for furthering the field of mathematics, and the more we can create spaces to celebrate a variety of ways of thinking, the better we all are.

Students have endless creative ideas, and we need to give them space to share them in class and in our assessments (because it's in assessments that the students often see what we truly value in our class). By valuing individual students' thinking and nudging all students to grow in ways that are appropriate for them, we are creating a more inclusive environment for all learners.

How might we proceed with assessment practices that honor students in this way? See Appendix D for sample unoppressive assessments.

ITERATIVE ASSESSMENTS

I developed the practice of iterative assessment in conjunction with a few of my colleagues and students as an additional assessment practice to the student portfolios I had been using. I wanted an individual accountability measure that looked for what students did know instead of what they didn't know. This process was integral for their learning to celebrate their mathematical thinking, and it trained my eyes to look at student work in a more asset-oriented way.

The premise of iterative assessments was that each Friday, for the first fifteen minutes, the students would sit individually and quietly at their desks with a handout that asked them to list everything they knew about a given concept. For statistics units, the prompt would include a small data set; for geometric assessments, it would include some images; and for functions units, it might include bare functions in function notation.

The first week the assessment was given was the baseline, where the students would share their thinking with respect to the prompt. They would sometimes just list questions, sometimes circle things they recognized, and, when possible, make mathematical claims about the prompt as given. For subsequent weeks, the same or a similar prompt was given, and students would share their growing knowledge of the concept over time. After four to five weeks, students could hold their papers side by side and visually show their growth in knowledge and understanding, sometimes even creating new mathematics out of the prompts given, thus being active participants in the creation of mathematics, not just becoming passive users of mathematics that other people did before them.

TIPS FOR DESIGNING A FRIDAY ASSESSMENT

- Phrase the prompt as "list everything you know or think you know about ...."

- Give students some guiding vocabulary to use.

- Make sure it relates to most of the mathematical goals for the unit.

- Prompt them to ask questions.

- Look for opportunities to get feedback on other elements of your class.

CHECK MARKS FOR "GRADING"

While written and verbal feedback would be ideal in all situations, I had to comprise a grading system of these assessments that allowed me to quickly acknowledge the students' thinking and quantify it in a way that helped show their growth. I developed a check mark system for these assessments, where each "mathematically correct" idea was given a check mark next to it. My class and I discussed what might make something "mathematically correct," deciding that it could be reasoned and justified and fit within the laws of mathematics as they currently existed. I let them know that I was not claiming to have all the authority on what was and wasn't mathematically correct. If the students ever thought I missed a check mark on the assessment, then I was more than happy to engage in a conversation and let them explain.

When I had time in my grading, I also tried to help students "ask the next question" or ask questions about their ideas that weren't yet mathematically sound to push their thinking. I also snapped pictures of some of their work to leverage their thinking and guide us into the coming week. Some of the images were celebratory of sound mathematical thinking, some were of questions students had asked, and others were of mathematical ideas that weren't fully fleshed out yet that might be unpacked during our upcoming week of work. In any case, the use of the assessments was intentionally fluid, nonpunitive, guiding, and celebratory of the beautiful mathematical thinking that students in the classroom exhibited daily.

SHOWING GROWTH WITHOUT CONSTANT COMPARISON AND OPPRESSION

Most of the time I followed this practice, I had the students store their assessments in a manila folder in the classroom. On the folder, they would track the number of check marks they got from week to week. Given that it was my job to assign them a grade that represented their learning and growth, we collectively decided that for "full credit" on the assignment, a growth of at least five mathematical ideas was prudent for each week (some years, students went with three) and that if the students had less than five, they could revise to reach the point where they would share five new ideas on the prompt. Usually by the third week of the iterative assessment, some students were writing over twenty mathematically correct and connected ideas on their papers.

Contrary to assessment systems that situate themselves as races to the right answers, in this system, the students with the most interest and excitement in mathematics take the longest and often run out of time because they are eager to make new connections. Students who are struggling are given sentence starters or questions as I walk around the room to get their brains ready to share their thinking.

ACHIEVING MATHEMATICAL GOALS FOR THE UNIT

Using assessment systems like this could potentially skirt grade-level-appropriate learning goals, a practice I am always in favor of. If students' mathematical ideas were separate from the mathematical goals of the unit, I would start a conversation with them, connecting their thoughts to activities from the week and seeking their understanding of our current mathematical goals on the prompt. This way, the pushing and nudging were grounded in their experiences and expertise, rather than being guided by me or an outside curriculum company telling them what they should know.

"WORKING THE SYSTEM"

Some students realized early on that if they started with fewer ideas, they would have less to do the following week to receive full "credit." What a wonderful opportunity for discussion when they would bring this up! The purpose of my assessments was to disrupt Freire's banking system of schooling so the notion of showing everything you know because you are proud of your learning (and not because you need a good grade) was paradigm-shifting and sometimes disorienting for students. I didn't expect them all to come along right away, so I normalized their hopes of "working the system" by inviting them to do just that (write nothing the first week and then around five ideas the second week, etc.). My suspicion (as I shared with them as well) was that they would not feel satisfied or proud of their work when they did this. This proved to be true for most students, who found a new source of internal motivation on these tests and a feeling of only competing against themselves and celebrating growth.

Another way of "working the system" that students found was to write repetitive information to receive a check mark for every entry on an in/out table or every point labeled on a coordinate plane. "These are mathematically correct; can I just list those?" they would ask. "Certainly!" I allowed. "If that is how you want to spend your time, you can certainly do it." Then, I would share a Friday Assessment of a ninth grader I once had who, instead of using her time to list one hundred inputs and outputs on a table, noticed new patterns on the table, including the rate of change and the second rate of change, and wrote a function to describe the rate of change. When I described her "discovery" of the derivative function to the student who had listed one hundred inputs and outputs, he agreed this might be a better use of his time. In this way, the notion of "correct answers" was democratized, and the power to have them was shared around the room. Additionally, the original celebration of having the most ideas gave way to the celebration of new ideas, new connections, and even new wonderings!

As education author Alfie Kohn says, "Grades create a preference for the easiest possible task. Impress upon students that what they're doing will count toward their grade, and their response will likely be to avoid taking any unnecessary intellectual risks."

FINAL COURSE GRADES

In high school, where I teach, it is common to hear students asking questions like "How many points is this worth?" or "Is there any extra credit to improve my grade?" These questions have always left me feeling like the educational experience I am giving my students is cheap and reductionist. My colleague Carlee Madis and I worked through this dilemma about eight years ago. We did some initial brainstorming and settled on a short set of hopes for our students for the coming year (see Image 5.1).

Reality of the Present:	Hopes for the Future:
• Students "learn" exclusively for a grade. • Students complete work because it is worth 10 or 20 or 150 points. • Students view teachers as authoritarians who judge students' rote content knowledge with scores. • Students form their identity exclusively around grades, particularly quizzes and tests.	• Students learn for gain in knowledge, skills, and self-improvement. • Students create extremely high-quality work because the work they do tells a story of who they are. • Students are viewed as collaborators within their learning journey. • Students have a broader scope through which to form their academic identity.

Image 5.1

We attempted to disrupt the structures of traditional grading systems to better achieve our hopes for our students' future. We recognized that it is disheartening to have a predefined self-worth as a math student based solely upon previous grades. Finding a way to help students have a positive view of their potential to be successful in mathematics is critical, and shifting our assessment practices and modifying our grading systems are integral to liberating students from the standard goals of grades.

To make this shift in our classrooms, we needed to be willing to allow the students to be central to the grading system, and we needed to view students as collaborators. It was necessary to invite students into the creation of expectations and, most importantly, to provide space for them to reflect and evaluate their own work and understanding. We kicked off the school year by sharing with students that we would not be assigning numerical grades on assignments throughout the school year. Alternatively, for every assignment, task, or assessment, students would be given our specific expectations, collaboratively set some of their own, and receive feedback in person and in our online gradebook (see Image 5.2). At the end of the semester, students would use evidence to propose a final letter grade through a mini-conference with us so that colleges could evaluate their level of work in the class (see Image 5.3).

Sample Student-Generated Scoring Rubric for Group Work Problems

EE (exceeds expectations)
- ☐ Student mathematical voice was evident
- ☐ The assignment had an idea about ALL questions (and an answer when necessary)
- ☐ Student explained ideas thoroughly AND made connections
- ☐ Student asked questions
- ☐ Student went above in one way or another

ME (meets expectations)
- ☐ Paper was complete
- ☐ Some thought process/work was present
- ☐ Some answers weren't thorough (just numbers)

NY (not yet)
- ☐ Only partially complete (some work done)
- ☐ Some answers present but lacked explanations
- ☐ No work shown

I (incomplete)
- ☐ Not turned in at all

Image 5.2

Guidelines for Grade Proposals

A range: Mostly ME, with a significant number of EE
B range: All ME (no I or NY)
C range: Some ME, some NY, few I
D range: A significant number of NY or I

Image 5.3

We finished the year with one-on-one meetings with each student to propose letter grades that would reflect their effort and growth from the year. In these meetings, the students reflected on their strengths and areas of growth, and then we shared observations with them. Instead of receiving a number score on a test as a simple indicator of mathematical worth, the students' identities were formed by the opportunities they took to use feedback to refine their work and persist through challenging problems.

While a shift in grading and assessment practices is critical in our quest to liberate students in our math classes, we would be remiss in this chapter on oppression if we didn't conclude by coming back to the ways that the content we study in mathematics can help us interrupt oppressive practices in mathematics, which is necessary for student-centered instructional practices. As noted by math education professor Rochelle Gutierrez, "Creative insubordination includes the following acts: decentering the achievement gap; questioning the forms of mathematics presented in schools; highlighting the humanity and uncertainty of mathematics; positioning students as authors of mathematics; and challenging deficit narratives about students of color." Books like *High School Mathematics Lessons to Explore, Understand, and Respond to Social Injustice* do a brilliant job of attending to some of these lessons, and we will share more of our explicit work in this category in the coming chapters.

SHIFTING THE CONVERSATION WITH PARENTS

One of my students' parents sent me an email asking, "Is she passing or not?" with a tone of aggression, and I wasn't quite sure whether the aggression was directed at the student or me. Despite clear language in the syllabus, regular emails with instructions on how we might make sense of our students' grades along the way, and the most recent email indicating grade proposals were coming soon and parents should be looking out, this parent just wanted to get down to the critical facts.

It is not surprising that even parents are quick to fall prey to the banking model of education. They want to know the basics: Is my kid passing? Are they an A student or a B student? If they are a C student, what consequences should I give them to motivate them to do better?

The grade proposal system brings parents into the conversation and humanizes and liberates everyone from the arrangement in which the teacher has the power to assign a grade that defines the student. I responded to this mom:

"Because my system of grading is built upon the student grade proposals, I can't say for certain yet. Can you please ask your child which evidence they are thinking of bringing in to justify their letter grade for the semester?"

"Oh yeah, sure," she replied, and then later, "I have looked at my daughter's work, and she showed me all the work she has done this semester. Some of it is beautiful, but a lot of it is still incomplete. We talked about how she needs to make some revisions, or her grade proposal isn't going to be something she will be proud of."

This was an ideal response for me! It opened a conversation between everyone on the student's learning team in a way that centered the student, their work, their thinking, and their brilliance, and de-centered the grade.

The process is slow, but it is necessary if we are to end the oppressive nature of assessment and grading in schools. It requires all stakeholders to be attentive and willing to come along for the journey. The stakes are high, but the alternatives are downright oppressive.

GIGI'S REFLECTION

Our previous chapter focused on the seemingly intrinsic intimidating nature of math, whereas this one concentrates on the everyday practices of that intimidation: the oppressive arms of arithmetic. "Liberating" is not a typical word used to describe math class. On the contrary, we can see how common oppression-based descriptors of math are in our collection of Dear Math letters. It is incredibly clear that the pressures of tests and grades do not incentivize but rather hinder mathematical curiosity and genuine understanding. Grades are viewed as threats instead of opportunities. Herein lies the main issue of oppression within our math schooling: grading is not something that students believe they have much control over. It is something done to them. Uprooting this feeling of inevitability is essential in the mission to achieve a more liberating classroom environment.

Throughout my years of project-based learning, particularly in high school, numerous exercises have exemplified this incendiary math mindset (Simon Bolivar would be proud)—the first being grade proposals. As simple as this practice may seem, allowing students to do their own analysis of their effort over the course of a semester, examine their work to use as evidence, and ultimately propose their grades based on their analysis and proof work is the antithesis of oppression. Completing this process (which I have done approximately eight times) has made me feel in control of not only my grades but my comprehension of the work of that semester.

The second practice was quite aptly named: Fri-yays! These once-a-week "quizzes" of sorts emphasized individual growth as opposed to rubric-based grading. Each Friday, I would come into class, and there would be a piece of paper with only one mathematical expression on it, waiting for my peers and me. This is not because Sarah vehemently loves to waste paper but because the open space was intended for us to fill it in. The only instruction given in tandem with our Fri-yay papers was to write as many observations as we could about the expression given. This could include graphs, tables, simplifications, connections, and even mere pieces of an idea. And for each correct observation I wrote, I would receive one point. The way to maintain an "A" on these iterative assignments was to get more points on your Fri-yay than you did the week before (kind of like a recursive function; that right there would totally be worth one point, for example).

These simple yet incredibly meaningful practices have helped to bring me out of my math-loving chrysalis and spread my liberating, symmetrical, math-loving wings.

INVITATION TO
REFLECT

- Have you ever experienced math in an oppressive way? If so, what did that feel like? If not, or additionally, whose voice might you listen to regularly to make sure you are considering the experiences of folks who are different from you?

- Which of your classroom practices are most oppressive to students? Which are most liberating?

CHAPTER 6

DEAR MATH, I AM FRAUDULENTLY FOND OF YOU

"Dear Math, I was told I was inherently 'good' at you. It was because of this, not a deep or genuine curiosity about your ways, that I became fraudulently fond of you."

STUDENT IN MY ninth grade math class was supposed to be working on a visual pattern task with her group. While two group members leaned in, shared ideas, and asked each other questions while pointing at each other's papers, this student postured herself with her paper on her lap and her body visibly removed from the group. She was invited into the group discussion a few times but ignored the invitations. After about fifteen minutes, she set her pencil down and looked up at me with a look that was somewhere between "I'm done," "What's next," and "This is boring!" I walked over and encouraged her to share her ideas with the group, but again she resisted.

As class wrapped up, I asked her if she wouldn't mind checking in on her math ideas at the end of class. She was eager to show me her work, which was beautiful, neat, and organized, and she did have a solution that was "correct" in many ways.

"Why didn't you want to share these amazing ideas with your groupmates?" I asked.

The student did not answer my question; rather, she jumped into a story about her math class the previous year. She said that her eighth grade teacher would give a quiz every day. The following morning, the teacher would start class by sharing the top students' scores on the quiz from the day before. This student was in the top three quiz scores every day and, even a year later, she was beaming to me about it. I listened to the story with interest, affirmed that she did, in fact, have lots of mathematical strengths and that I was not surprised she did well on her teacher's daily quizzes. When I re-asked my question about working with her group, the student's face twisted into what I can only describe as disgust. "Why would I share my ideas with them? They can come up with their own ideas."

I tried to keep a straight face as I looked back at her. "Because we are better when we share our thinking with others and listen to their thinking. Our community is better, and you are better!"

"Well, I just want to finish my work as fast as I can so I can move on," she responded.

"Where do you want to move on to?" I asked.

"Whatever I need to learn next," she responded quickly.

I could tell this conversation was torture for her, so I thanked her for sharing her story with me and sent her to her next class. This conversation was the beginning of a long discourse between this student and me throughout the year. My goal was for this student to develop a love for mathematics. What she currently had, however, was more like an addiction derived from being called the "best" day after day in her eighth grade class.

FRAUDULENT FONDNESS IN THE CLASSROOM

From a young age, Gigi and I both loved math as kids. This love of math came from an interest in the patterns and relationships with numbers, but it was also cemented by teachers and standardized tests that regularly told us that, due to our speed, accuracy, and ability to follow directions, we were strong mathematicians. Having our innate skills in mathematics reinforced by these systems and based on this false definition of "mathematical" was ultimately damaging. Thanks to Gigi's written letter, we have come to describe this adoration for the subject as a "fraudulent fondness." Fraudulent fondness speaks to the experience of many traditionally "high-achieving" math students who don't necessarily love math but feel drawn to the public acknowledgment of being good at something. They are fond of the feeling of being good at mathematics but not actually fond of the beauty and connectedness that mathematics offers.

The type of praise and affirmation that usually results in fraudulent fondness can often create what psychologist and author Carol Dweck describes as a "fixed mindset." A student's math identity is grounded in being fast or right, both of which are at odds with the broad, creative, sense-making subject that math really is. The even more unfortunate piece to this is that some students who aren't as fast or as right become marginalized from the math community. We will speak to the experiences of those students in a later chapter. For now, let's fix our attention on the fraudulent fondness that some students come to develop for math and discuss interventions to help foster a more genuine, whole, and deep love for the discipline.

If a student is used to being told they are good at math because they get the highest test grades, finish the quiz the fastest, or, as Gigi mentioned, maintain a "simple safety net by my guaranteed A," coming to form a math identity based on other qualities can be hard. The high-stakes testing and reporting in most math classes are grounded in the same descriptors that students are accustomed to. It will take a sustained effort by the teacher and the willingness of the student to begin approaching math class with an open mind, ready to push on the cultural norms of math.

"Dear Math, Starting as early as elementary school, I enjoyed your company. This was mainly due to the fact that I was told I was inherently 'good' at you. It was because of this, not a deep or genuine curiosity about your ways, that I became fraudulently fond of you. My fondness, though, was rather duplicitous, for I found you simple, and while commonly envied, this false conception of simplicity is what stopped me from loving you. It's what stopped me from viewing you with interest or inquisitiveness." –Gigi, tenth grade

TESTING, RANKING, AND SORTING

The nature of traditionally scored tests and quizzes, particularly when the reporting is public and numeric, gives an oversimplified view of what it means to be mathematical. Students get information in the moment about whether they are or aren't mathematical, and teachers look at the results of these tests and quizzes and quickly rank and sort students by perceived skill level. While assessing students and making sense of their skills and the learning goals of lessons are integral to teaching, the use of this information to rank, sort, and make publicly known who was "the best" results in students developing a fraudulent fondness. Even teachers with the best intentions, when handed lists of their students ranked by a given score, tend toward the condensed thinking of "high kids" and "low kids." Students are such complex humans that seeing lists like this feels like it can help us instructors make more sense of the chaos. However helpful they seem, structures that make quick judgments based on a narrow set of skills tend to hurt our perspectives of the mathematicians in our care.

As Jo Boaler shared in *Limitless Mind*, "Mathematics can be a beautiful subject of ideas and connections that can be encountered conceptually and creatively, but students who attend schools that treat mathematics as the memorization of procedures, and that value the memorizers who can speedily regurgitate what they have memorized, turn slow, deep thinkers away from the subject." The education system has consistently put prolific memorizers on a pedestal (backed by standardized testing), but this does not accurately reflect a true understanding of mathematics. It is imperative that we give students a new way to describe themselves as mathematicians and gauge their learning on their terms.

"Dear Math, You used to be so easy to understand back in elementary school, like 2 + 2 = 4 or 4 x 7 = 28 and all that. I mastered you like you were nothing. I memorized your multiplication table, which was my biggest challenge back in elementary school, but it was worth it because I got a pass to go to the pizza party that was being held for the students who could memorize all of the multiplication table." –Edward, tenth grade

TIMED MULTIPLICATION TESTS

Many of the fractures we read about in students' Dear Math letters occured in third or fourth grade when students were made to feel that they weren't good at math because they weren't the top "performers" on timed multiplication tests. Conversely, as we see in this chapter, students develop a fondness for being the speediest mathematician because they are made to believe they are "the best." In reality, they are good at memorizing and regurgitating— a skill that, at best, is a small part of being a mathematician. From such an early age, students can come to believe that being mathematical is based on this one skill. Even if other mathematical opportunities are given to the class, the experiences that are easiest to recall and categorize are prioritized in students' stories of what impacted their identity development.

Timed multiplication tests, and timed tests in general, are a practice founded on past assumptions about learning. Based on how frequently they show up in students' Dear Math stories as a foundation of either fraudulent fondness (I was the best) or exclusion (that's when I knew I wasn't good at math), it is abundantly clear that this practice must end.

"Dear Math, I have felt successful in math. In elementary school, we took our '12s multiplication table test.' I had been studying for it really hard with my brother, and when it came time to take the test, I felt prepared, a feeling that is often rare for me to have in a math class. As I took the test, I was flying through the questions. I turned my test over, pencil down, right as it hit the thirty-second mark, knowing I had forty-five seconds to complete every question as best I could. Mrs. Brennen picked up my test, reviewed it for a couple minutes, and called me to the back of the classroom to tell me the news that I had, in fact, completed every problem and correctly! I was then promoted to the "prime math students" math group. ... The pressure was high from a young age to impress not only my teacher but my peers. Although it curated good pressure to study, it was also nerve-racking and embarrassing to even have the thought of failure cross my mind." –Scout, tenth grade

HIGH-STAKES STANDARDIZED TESTS

I recall a story from a colleague who spent a year with a student who had traditionally struggled in math. The teacher had cultivated deep learning, growth, and the creation of beautiful work in the student, who came to see themself as a mathematician for the first time in their life as a tenth grader. This student then took their state-mandated tests in the spring, received the results in the summer, and declared that, apparently, they hadn't learned anything after all.

Daniel Koretz, an expert on educational assessment and testing policy, states that excessive high-stakes testing undermines the goals of instruction and meaningful learning. While this applies to students who perform poorly on tests, it similarly impacts students who perform well who might come to see the results of that test as the end and ultimate success of their learning. This is the fraudulent fondness we speak of: the notion that students decide that they are or aren't mathematical based on a slim set of measures.

TALKING AND NOT TALKING

Defining the condition of being mathematical on a slim set of constraints is not unique to tests and quizzes. I recall a time in my teaching career when I decided my main goal was that students would speak more in class, letting them take the mathematical authority in the room. I started a system where students would lead our warm-up discussions while I would sit in the back of the room and track who talked. Students would accumulate tallies throughout the week, and this became part of their classroom participation self-assessment at the end of the week. The trouble with this system was that it relied only on the notion that students talked or didn't talk. Once again, a slim set of measures by which to succeed and be part of the mathematical community. I could neither appreciate the insights students might have had that were not shared aloud nor see that some of the shared ideas were not mathematically significant or thoughtful. In fact, some students who decided to work the system started talking for the sake of talking! The fondness here was for talking and getting points, as opposed to growing as a mathematician or bettering the mathematical community.

FOSTER A GENUINE FONDNESS FOR MATH

The combat of fraudulent fondness must begin with new anchors for fondness. Math teachers must provide students with a more complete and accurate version of what it means to be mathematical and expose them to a new language to describe themselves as mathematicians. This takes intentionality and practice.

The word "smart" is often tossed around and is rarely clearly defined. I hear it most commonly referenced when someone knows a complicated-sounding name for something, answers a seemingly complex question quickly, or recalls trivial information with ease. Many students are declared "smart" based on their grades on tests and quizzes or in courses overall.

> "Dear Math, As I grew up, you did too; you grew more complicated. It took me longer to analyze your different questions and asks, but I never gave up. I was so attracted to that feeling I got when I was able to get a correct answer and show off, so I always kept trying, and it seemed to pay off." –Dania, eleventh grade

SHIFTING AWAY FROM PERFORMING
AND ANSWERING

Math teachers have often been trained through their years of schooling and then years in the classroom to listen for a correct answer and applaud that. It takes a great deal of restraint to keep a straight face and not share a quick "right" or "wrong" when a student responds. Instead, when students share an idea or strategy, try these responses that maintain neutrality and inquisitiveness:

- "Thank you for sharing your strategy."
- "Can you share more about this particular part of your strategy?"

- "What questions do we have for this person?"
- "_____, will you please restate this thought for us?"

The shift away from a focus on right and wrong answers in class helps prevent students from building a fraudulent fondness. As students see that their ideas are valued and received by the classroom as worthy of discussion, they begin their journey toward a fondness grounded in creativity, sense-making, and collaborative problem-solving.

In her book *Strength in Numbers: Collaborative Learning in Secondary Mathematics*, Ilana Horn writes: "Judgments about who is smart based on prior achievement or social categories violate a fundamental principle of equity and are consequential: learning is not the same as achievement" (Horn 2012). Here we see that learning and achievement must necessarily be separated in cases where achievement is grounded on a narrow set of descriptors (as is often the case for standardized tests or even most unit tests given at district levels). Furthermore, achievement is one of the foundations, as described by students in their Dear Math letters, where fraudulent fondness manifests itself.

How might we begin the journey of redefining intelligence for our students in service of both real learning and a more genuine fondness for math? We might journal about how we think about the term "smart" and reflect on the ways that their understanding of the word is steeped in achievement as opposed to learning and a broad set of skills. We can then push ourselves to think about other ways students might exhibit skills in the classroom and then keep a watchful eye on this. In my work and in the work of teachers around me, I have seen this practice result in:

- an increased ability to notice and name ways students are demonstrating critical thinking in math class

- greater ease in finding language to give feedback about "smart" things students are doing

- incorporating an expansive definition of "smart" into planning, such as thinking about whether activities, assessment, and grading are valuing diverse ways to be smart in math class

I call this reflective work "the retraining of our eyes and ears" in our math classrooms. What are we looking for and listening for? See Image 6.1 for a list of mathematical attributes I look for in my students.

- Coming up with and testing many strategies, even if some fail
- Visualizing complex objects and/or representing things visually
- Finding background information by reading, researching, and asking others
- Noticing and describing patterns
- Representing things using symbols, numbers, and mathematical notation
- Clearly sharing ideas and thought processes with others
- Making connections between different ideas and/or seeing "the big picture"
- Asking questions that challenge or build on ideas of others
- Creating simple examples that help build understanding

Image 6.1

We use this list to "bring our strengths to the table" when entering a groupwork situation and to reflect on our work at the end of a classroom experience.

The Common Core Standards for mathematical practice offer helpful language for actions we hope to see and call out in our students:

1. Make sense of problems and persevere in solving them

2. Reason abstractly and quantitatively

3. Construct viable arguments and critique the reasoning of others

4. Model with mathematics

5. Use appropriate tools strategically

6. Attend to precision

7. Look for and make use of structure

8. Look for and express regularity in repeated reasoning

When students gain language to describe a broader set of mathematical skills, habits, and practices, they start to describe themselves as mathematicians more readily and wholly. They can verbalize ways that they are strong mathematicians and ways that they are growing as mathematicians. They can see that everyone in the room has strengths and evidence of growth. (For more about helping students build strong habits of learning, see the book *Hacking Student Learning Habits: 9 Ways to Foster Resilient Learners and Assess the Process, Not the Outcome* by Elizabeth Jorgensen.)

ASSIGNING COMPETENCE

It's critical for teachers to share language for how we describe mathematics in the classroom and how we utilize the notes. A colleague of mine, Chris Nho, calls these "professional mathematician moves." When a student attends to precision or notices a repeated pattern, Nho says, "Wow! That's a professional mathematician move right there." The students grow in their understanding of what mathematicians do and the various ways to be mathematical. Research professor Deborah Ball defines "assigning competence" as a set of practices that:

- broaden and label what it means to be competent in mathematics
- intervene to position who and what is seen as competent in math class
- support individual students to develop their mathematical and academic identities

Assigning students competence will help them in taking more ownership of mathematics. Fondness for the discipline is decidedly different from the addiction to glory and achievement that we describe as a "fraudulent fondness." This may look like a willingness to take risks by trying challenging math or asking questions, demonstrating a greater desire to listen to other students, and showing a stronger growth mindset around mathematics learning. The strategic use of public feedback impacts status dynamics in the classroom. The teacher considers what they know about each student's academic and social status to decide when to intervene. As Ilana Horn states in her book *Strength in Numbers*, "Assigning competence is a form of praise where teachers catch students being smart" (2012). She reminds us that as we give feedback to a student, explaining how their action supported mathematical learning for themself or their classmates, we must make sure that we give feedback that is public (not just to the individual student), specific (communicates what you value), and intellectual (highlights how their contribution helped move the math learning forward). For more on assigning competence, please also read Elizabeth Cohen's *Designing Groupwork*.

SCAFFOLDING STUDENTS TO CALL OUT MATHEMATICAL STRENGTHS IN OTHERS

As students begin to see a previously low-math-status peer as mathematically competent, they will listen and respond to that student's mathematical ideas more often. That student will be invited into the mathematical conversation, their learning will be deeper, and their identity will be grounded in more positive stories of contributing to a challenging problem with a group.

As a teacher, you have only some of the power in the room to call out what it means to be mathematical. The students also have a lot of power to call out mathematical strengths in one another. One of my favorite practices when students enter a new group (we switch to new randomly assigned seats every two weeks) is to ensure that everyone is coming to the group with strengths and areas of growth. A former colleague of mine, Bryan Meyer, shared with me the idea of keeping strips of paper in my room, each with one of nine attributes of a mathematician, and when it's switching-seats day, students come up to the front, grab the slip that describes their strength, and sit down at their new table. In this way, students are literally "bringing their strength to the table." As they reflect on their group work that day, they can discuss the ways that people in their group exhibit those strengths.

In her letter to math, Gigi mentioned, "I found you simple, and while commonly envied, this false conception of simplicity is what stopped me from loving you. It's what stopped me from viewing you with interest or inquisitiveness." Without language to describe the mathematics that is happening in the community, words like "simple" or "easy" become the goal for students. When we provide a greater diversity of skills, habits, and language for our students and model them daily, students can find a fondness for mathematics that is real, not fraudulent.

GENUINE FONDNESS IN THE CLASSROOM

In 2019, I had the opportunity to take part in a public lesson study. This means a team of teachers come together and design a lesson over eight weeks with the purpose of teaching the lesson to a group of students while being observed by a crowd of parents and community members. The lesson is then debriefed by the lesson study team, and a few knowledgeable others comment on the equity goals and the content goals of the lesson.

My lesson came at a time when my students were partway through a unit on quadratics, and we were exploring types of quadratic functions that couldn't be factored. I had taken the students outside the day before to toss some balls up in the air and do some reasoning about the vertex. During the trip outside, I captured a photo of two students tossing their balls side by side, and that became the launch of the task the next day: to reason about whose ball had gone higher. One student's toss was documented by a graph, which the class collectively made sense of in our warm-up. The second student's toss was given as a function in standard form. Of course, I had approximated the function. The reveal of this equation brought curious frowns, and the students had to decide, given two representations, whose throw went the highest. While not situated in a purely authentic modeling context, the lesson study team and I had created an intellectual need to solve a problem (a throwing competition between two peers) that everyone was interested in solving.

As I moved around the room, students worked in groups to unpack the problem and make sense of the scenario with the limited information they were given. Six students shared strategies on the board for solving the problem. Even though we didn't reach a complete solution that day, we left with lots of potential conjectures for whose ball went the highest (and the vertex form of a quadratic).

At one table was a student named Lou who had been excluded

from the mathematical community for much of his life. During individual think time, he had organized the quadratic equation to highlight the ways it could not be factored. He had used the "box method" and was trying to figure out what would go on each of the outside parts to make the equation factorable. "What do we need to do to make this be able to be factored?" he asked. This was a critical question for both his table and the class. His particular representation of the quadratic and his question became paramount in his group's ability to think about the problem in a new way.

The exit ticket that the lesson study team had designed asked, "What is something you learned from someone else today?" In their exit tickets, every person at Lou's table shared that Lou's idea was what propelled them to greater success.

The next day, I asked Lou to stay after class for a few minutes, and I shared some of the exit ticket language with him. He tried to brush aside the compliments, but I reaffirmed that he had led his group mathematically the day before and that it was, in fact, critical that he was a part of their group yesterday. He claimed that he hadn't answered the problem correctly, so he didn't feel like he had been helpful. I named his contribution, however, as having "visualized a complex equation in a new way" and told him that without this work, nobody would have been able to solve the problem that day. Everyone in the class has an important role to play on different days, and I was grateful for Lou's mathematical brilliance that day. I applaud his peers for having taken notice and naming it as well.

GIGI'S REFLECTION

As a young student who rather quickly grasped the elementary math topics being taught in classrooms and whose mathematical prowess included speed, my competence as a mathematician was never questioned. I would look down at the bright red A+ on my math assignments and become intoxicated with the validation, a feeling I would chase for years without ever questioning why my skills were celebrated while my classmates' were critiqued. For me, math was never a team sport. It was like tennis. My opponent was the expression/equation/problem at hand, and I was like Serena Williams—globally acclaimed for her excellence in the game, winning streak, and great arms. When I came up against Martina Multiplication or Steffi Statistics, all that mattered was getting the ball over the net and winning. Every time I'd slam down a serve and get the answer right, the crowd would go wild, I'd receive a gold trophy with a bright red A+ stamped on it, and I'd once again feel intoxicated.

It wasn't until high school that I realized math wasn't a tennis match. This was when my math teacher meddled with my "win, win, win" mentality. This was also when I learned that not everyone had the same positive relationship that I had and that however much it meant to me to get that A+, it meant way more to the kid next to me who didn't get one.

I discovered, through many trials and tribulations, that my juvenile intoxication with the exclusive definition of success was just that: juvenile. It's not an easy epiphany to reach when you're one of the people standing within the confines of the velvet rope, and in fact, it took me being left on the outside to realize the problem. This fraudulent fondness of mine stemmed from my misplaced value in correctness as opposed to inquisitiveness. Ease over struggle. Because of this, I put all my focus into ensuring math was easy for me—ever striving for that simple, gleaming A. With this mindset and years of repetition, coupled with consistent praise from my teachers and peers, I felt highly confident in math class. Thus, when one of the rare math tests of middle school was announced, I didn't experience much of the dread, nervousness, or anger that many of my classmates felt. Math was easy, so of course, I was confident. Until it wasn't. I failed that particular test four times. I was kicked out of the cushy, privileged VIP section and left on the outside of the red velvet ropes, longingly staring in.

I was no longer fraudulently fond of math. Rather, I hated it, plain and (ironically) simple. This was before Sarah Strong became my math teacher. It was as though I was a seventeenth-century flat-Earth believer, and Sarah was Galileo. She taught me that math was anything but simple. I learned the depth of its true importance and that math was truly everywhere. I learned that my perception of there being only one answer to each equation or expression was entirely false; furthermore, that in some ways, the solution was far less important than the process of reaching it. For the first time, I was challenged by math in an appropriate way, and it stimulated my mathematical thinking more than ever.

INVITATION TO
REFLECT

- What is something that you have a "fraudulent fondness" for?

- What indications might you look for that your students are developing a more genuine fondness for math?

CHAPTER 7

DEAR MATH, YOU ARE BEAUTIFUL

"Dear Math, Although you've been difficult, I see the beauty of you."

IT WAS EXHIBITION night, and the excitement in the air made the space feel light and celebratory. Despite the crowds and the variety of options for visitors, there was a sense of purpose, direction, and anticipation as people moved around the building, eagerly seeking what was to come next. Parents, students, and community members were bustling about, heading toward different classrooms where students had set up displays of their work to share their products and processes from a semester's worth of learning.

At our school, exhibition night is the pinnacle of the entire year. It is the night when every classroom transforms into an interactive museum. In my tenth grade math classroom, we had spent the last eight weeks in our Shape Shifters math project, which I designed with my colleague Dylan Bier, exploring ideas of similarity, right triangle trig, and solid geometry. The students had created a couple of different final products, but most stunning were their shadow art pieces. Shadow art is a form of sculptural art where the 2D shadows are cast by a 3D sculpture. The students spent time searching the internet for examples of shadow art. The websites of famous shadow artists like Vincent Bal, Fabrizio Corneli, and Kumi Yamashita were a starting point for the work. The students then explored the relationships between the sides of similar triangles, and the trig ratios emerged. Finally, the students analyzed shadow art installations by making sense of the light source, the object, and the shadow.

As the students created their shadow art pieces, they went back and forth between trial and error and the trigonometric underpinnings of the relationship. They drafted, they provided feedback, they experimented, and all the while, we all started paying extra special attention to the shadows around us and, metaphorically, the shadows of our math stories that needed to be brought into the light. For exhibition night, the students had set up their shadow art pieces to be interactive. All the visitors were supposed to stand in a set spot and hold their cell phone flashlights at a certain angle so the shadows were cast in just the right direction to form an image. Parents usually love exhibition night, but for this particular exhibition, I got more parent follow-up emails than usual, thanking me for creating such a beautiful, delightful, and interactive presentation. Even though the math wasn't used in a way that modeled a real situation, the parents saw the beauty of math in a way they hadn't seen before, and we were delighted!

"Dear Math, I know our future will hold many positives and negatives—a sporadic nonlinear splatter of points on the graph of my life and our relationship. I love you." —Kieran

BEAUTY RESEARCH

The term "beautiful" isn't the most frequently used term in student letters, but it shows up periodically, both explicitly and implicitly. Beauty refers to a combination of qualities, such as shape, color, or form, which pleases the aesthetic senses, especially sight. Some student letters mention a visual beauty in math, particularly when they reflect on symmetry in art pieces and final art products, but few reference this visual type of beauty when speaking about their math experiences as a whole. Rather, beauty more frequently shows up in student letters in the feeling of a "beautiful experience." Many students describe this as satisfaction because of how things work together. Quotes like "You make us the best we can possibly be," and "You are a puzzle I was excited to solve," and "I love you" all speak to cerebral feelings of a sense of beauty that a particular piece of math learning made them feel.

Francis Su references four types of beauty in his book *Mathematics for Human Flourishing*:

1. sensory beauty

2. insightful beauty

3. wondrous beauty

4. transcendent beauty

According to Su, *sensory beauty* is the beauty of a patterned object that can be natural, artificial, or virtual. His examples include fractals, Islamic art, and sound wave patterns. This is reflected in students' acute stories of beauty. *Insightful beauty* is the beauty of understanding, elegance, or a feeling like you stumbled onto something. Puzzles sometimes have insightful beauty. *Wondrous beauty* is the beauty that emerges from feeling wonder, as when you cannot help but ask why. Some math equations can be admired without understanding them. And *transcendent beauty* arises when one moves to a greater truth of some kind. When you experience this, you feel a profound sense of awe. Su's brilliant definitions and descriptions helped me better understand the themes I was seeing in my student letters.

CONSOLIDATION AS A MEANS TO BEAUTY

Consolidation techniques help students make connections and synthesize their learning, and they can also create an avenue toward more beautiful work and beautiful meaning-making in mathematics. Let's dive into the why and how of consolidating *beautifully*.

It is common to hear students say, "I forgot that," or "I never learned that," or "I just don't get it." These phrases are often an indication that their learning hasn't yet been consolidated. This lack of interconnection hinders their ability to see the content they are learning as either an intriguing new insight or a transcendently beautiful concept. To achieve opportunities to see beauty in math, we need to give students ample time to learn deeply, including consolidating their learning. Consolidation is the process by which students learn an idea they can later retrieve and use. In their book *Brain-Based Learning*, Eric Jensen and Liesl McConchie cite a need for a thoughtful use of time in student lessons and learning. They claim, "Time should be a significant consideration in the process of fostering memories. Essentially our brains need some time to decide "keep this" or "trash that" for much of school-based learning, the hippocampus works on a timed schedule, usually just days when the new experiences are undergoing memory consolidation." When we provide students with the opportunity to properly keep or trash an idea, we allow them to experience that feeling of solving a puzzle or making a new connection between ideas. It gives students a chance to see how wonderfully math works.

Akihiko Takahashi and his education researchers in Japan give us further insight into this idea with the notion of matome. In his book *Teaching Mathematics through Problem Solving*, Takahashi notes that "Matome usually consists of two parts: first, students look back and sum up what they learned that day in class, then they write it down in their notebooks along with their reflections." As students write reflections and sum up lessons in their own words, the teacher can see what each student is taking from the lesson and how they felt about the lesson that day.

"Dear Math, We just have to keep trying till we get it right. There's no need to rush; we just need to stay on the same page to really connect. I know that with your help, we can go far in life. It will get very challenging, so that's when we need to take some time apart. I know that we will be friends forever, we'll help each other out, and we won't give up on each other. In the future, I hope we will still get along and everyone keeps their mindset steady. Everyone should be able to see what you're valuable in and see your beauty and how wonderfully you work." –Seventh grader

CULTIVATING MORE BEAUTY IN MATH CLASS

While matome, consolidation, and transfer techniques help to cultivate beautiful lessons and daily learning, another facet to beauty is the creation of beautiful work over the course of a unit or project. My school takes a page out of Ron Berger's book *An Ethic of Excellence* by advocating for students to create "beautiful work." Throughout a unit or project, students are in a continual process of creating and iterating work that is growing in detail, complexity, and connectedness. Throughout this process, it becomes increasingly beautiful and represents their learning. Berger says, "I want a classroom full of craftsmen. I want students whose work is strong and accurate and beautiful. Students who are proud of what they do, proud of how they respect both themselves and others."

While sometimes this quote is taken to represent pieces of woodworking or artwork, it also applies to various other products that students might create that show new connections between content and the students' lives. It can also show a beautiful story about how the student has learned and grown. Berger also says, "I use the phrase beautiful work broadly ... work of excellence in any discipline is beautiful to me, and I don't hesitate to label it so." I find beautiful work can be continually redefined by the community as they look at previous beautiful work and build on it.

Here are ideas for different methods of consolidating learning at the end of a discussion and finding the beauty in integrating new ideas:

Journaling

- Ask students to write down the big ideas from their learning that day and how the ideas connect to the main lesson from the day before.
- Pair share.

Anchor charts

- Ask for two or three volunteers to share. The class discusses differences and similarities in the takeaways, and then each student revises their own summaries.
- Students share out and the teacher (or a student) captures the ideas on a large poster paper to be displayed in the classroom.

Titling the lesson

- Students can come up with a title for the lesson, and it can be written on the board, anchor chart, or notebooks at the end of class.
- Share titles for the lesson out loud and vote on the favorite.

As documented by Elizabeth Green in *Building a Better Teacher*, "Title writing helped tie the lesson together." A summary like "Times 2 and divided by 2 are brothers!" was a young boy's selection for the title of a lesson focused on showing the relationship between multiplying and dividing. Both this young boy and the students in that class undoubtedly felt feelings of beauty as they participated in these types of activities where they were able to creatively express feelings about their learning.

In the math classroom environment that I was striving to create alongside my students, this work happened through portfolios and summaries of learning.

> "Dear Math, You are challenging, but I like challenges. You are there when there is something I have to figure out. You have been there for me my whole life. In third grade, you were my favorite subject. If it wasn't for you, I wouldn't have met my amazing teacher Ms. Frost. ... You will be there when I am all grown up and trying to get a job."

PORTFOLIOS

As defined in Berger's *Leaders of Their Own Learning*, a portfolio is "a selected body of student work—with reflections—that provides evidence of a student's progress toward standards, learning, targets, and other character growth." In addition to the selection of work, the portfolio system uses student reflection as further evidence of growth. In using reflection, students are given the opportunity to share their voices and, perhaps even more importantly, to return to something they worked on earlier in the week, month, or year. Another important part of reflection is summarizing content learning from each unit so that students can compress the ideas they explored.

The purpose of a portfolio assessment system is to give a large part of the assessment power to the student. This shows they are valued as an authority on their learning and they can articulate and accurately describe how well they are meeting the goals because they are the ones who did the work. Students can compare their work to the high-quality examples they were given and make plans for the next steps.

PORTFOLIO INSTRUCTIONS

Each week, the students complete math tasks and activities. At the end of the week, they reflect on a math concept they learned, how they grew as collaborators, and how they progressed toward their specific goals. Then, they store these in their binders, and we record them on a shared class table of contents.

It is important to set up routines for portfolios at the beginning of the unit to make sure students are keeping their work in an organized place where they can return when they need to compile their portfolios. This can also be digital. A digital portfolio can be especially useful because students can maintain it throughout their educational careers.

In class, we create a rubric together. Students brainstorm and discuss the works that need to be evident in the portfolio for them to justify an A. They then use this as a checklist for completing the portfolio and critiquing each other's work before submitting it.

Feedback on portfolios is broken into three parts: citizenship—whether the work is completed, on time, and inclusive of links to evidence; content—whether the students have provided accurate demonstrations of growth in understanding mathematical concepts; and self-reflection—whether students are honest and thoughtful about their overall growth in that unit.

LEARNING SUMMARIES

At the end of a unit, students summarize their understanding of that unit (called Summary of Learning) and justify the grade they feel they have earned. This summary must be supported by the remainder of the portfolio and requires students to go back and look through their feedback and quality of work. Students can revise their work and improve their grade by responding to my suggestions. The students create a visual representation of their learning that summarizes and shows connections between the ideas in the unit, potentially connecting them to the project as a whole. Before I go into more of our work to create these, see Images 7.1, 7.2, and 7.3 for samples of student work.

Image 7.1: Student work by Gigi Butterfield.

Image 7.2: Student work by Cloe Moreno.

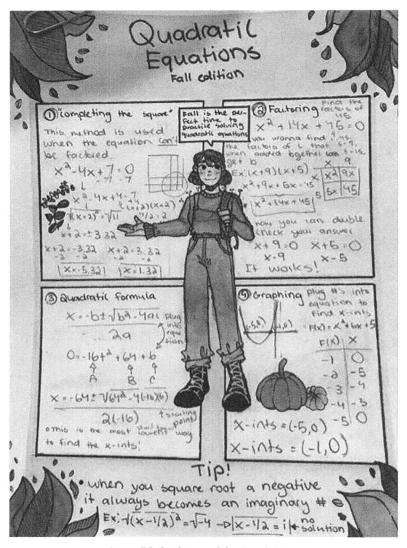

Image 7.3: Student work by Cara Salazar.

To continually strive toward beautiful work in our summaries of learning, my class looked at samples and brainstormed about the facets of a "beautiful" summary of learning. The students came up with the following list of attributes.

WHAT MAKES A BEAUTIFUL SUMMARY OF LEARNING?

- a sense of exploration/deep thinking
- simplicity/clarity
- organization
- examples of what you're writing about
- clear thoughts and ideas
- clear effort
- explanations of all the most important topics
- creativity/originality
- personalization relating to your learning style
- visuals/graphs paired with descriptions

PORTFOLIO REFLECTIONS AS A FORM OF MATOME

I've found that students learn more when they work with an idea, let it get cold, and then return to it. Portfolios and summaries of learning offer this opportunity to step back before reflecting because in the cycle of compiling portfolios, students return to their work throughout the unit or project, often revise it to make it better, and then metacognitively discuss the entire process with given prompts.

General self-reflective questions might include:

- How has your thinking changed?
- What were your significant learnings?
- What have you realized about how you learn?

More specific self-reflective questions might include:

- In what ways did you have to show perseverance?
- How did you solve the problem, and why did you feel it was an effective way of doing so?
- Recount the work you completed with your peers. How did you contribute to the work? How did you contribute to the group's process?

We need to continually give students the chance to consolidate, reflect, transfer, and then create work they are proud of. In the article "What is Mathematical Beauty?" Jo Boaler states, "We are now in the twenty-first century. Students do not need to be trained to be calculators—we have technology for this—but they do need to experience mathematics as a beautiful and connected subject of enduring big ideas. Students who learn through big ideas and connections enjoy mathematics more, understand mathematics more deeply, and are better prepared to tackle the big complex problems and discoveries that they will meet in their lives." Portfolios and summaries of learning help our students discover their capacity to create mathematical beauty. For a sample portfolio requirements sheet, see Appendix E.

NEW BEAUTY CONVERT

I'd like to share a story about a mathematical beauty convert in my classroom. Reflections, rotations, and translations had become a regular part of our rhetoric over the previous four weeks. We had explored congruent shapes, conjectured about how we might prove congruence, and, from a teacher's vantage point, accomplished our learning goals for the unit. The time had come to begin assembling our portfolios in which the students would reflect on their learning and create a summary of learning to outline what they had gained from the past unit.

Yosef had spent the unit quietly doing his work, minimally contributing to groups, and rarely sharing his work with his group despite my attempted interventions. I could see his mathematical brilliance was within him, but I don't think he saw it in himself, nor did his peers yet see him as mathematically brilliant.

When the day came to turn in portfolios, Yosef walked in carrying a packet of papers neatly fastened with a black binder clip. At the front of the portfolio was his summary of learning. When my eyes focused on the page, my jaw dropped. The artistry was clever and thought-provoking, the mathematics was precise, and, most importantly, it was so *him*! I had only seen the first draft of this work in class because he had wanted to work on it more from home, so I was completely unprepared for the level of beauty on the page.

Knowing that Yosef was shy, I tried not to get overly excited and show the page to everyone. Instead, I said, "Wow, that's beautiful! Can you tell me about your process of making this?" Yosef described how, when he thought of rotations, reflections, and translations, he thought of birds moving through the air. He loved drawing birds, so he wanted to show his mathematical learning in this form. His drawing outlined three different types of birds, one whose wings were showing reflective symmetry, one that was replicated right next to it showing a translation, and a third that was flying through the air, rotating as it went. Each bird was labeled with a notation to show the parts that were congruent.

In our gallery walk that day, many of Yosef's peers showered him with praise for his summary of learning, and it has lived on my desk for years as a reminder of the beautiful work that students can create when we give them space and permission to express the beauty of math as they see it.

"Dear Math, Math is definitely a puzzle. A puzzle I was always determined to solve. You've given me some of the best supporters and the biggest "aha!" moments. Never again will I think of you as terrible, never again will I think of you as boring, and never, ever again will I think of you as simple, for you are anything but, in the most amazing way possible." —Gigi

GIGI'S REFLECTION

Math is somewhat akin to the cliché girl-next-door. You barely notice her every day, but when she lets down her hair and takes off her glasses, suddenly, she's beautiful! Obvious unrelatability and chauvinism of that trope aside, the metaphor still rings true. Math is intrinsically beautiful. It exists within humanity's greatest beauties, such as music and architecture, and in nature's fractals and concentric circles. Sadly, though, contemporary teachings and perceptions of math require us to view it with a practiced lens to see it as such. Truly passionate and well-informed educators must do the work of letting down math's hair to inspire the student to appreciate her profound beauty.

There is beauty at every level of mathematics. One beautiful aspect that is frequently discounted is the holistic process of learning arithmetic. The magical "aha" moment when the final part of a math concept clicks is a spectacle whose allure rivals The Colosseum or Grand Canyon. As someone who has experienced it from both positions, the magic is tangible for the student and the teacher.

The understanding at the root of the work is as beautiful as the work itself. There is an inherent beauty in the rawest, most embryonic actualizations of math, such as a sheet of lined paper with graphite trials and errors scribbled across it. The pure excitement of mathematical inquiry in the meager confines of pages of chicken-scratch is an undeniably magnificent presentation of curiosity. A matured, tangible form of math learning is similarly magnificent. In place of finals in Sarah's class, we curated summaries of learning. These were aptly named assignments, allowing us to summarize all we had learned throughout the course of a given unit. It was exciting to have no constraints on the ways in which we executed our summarizing. Students who excelled at artistic programs used Photoshop and Illustrator to create beautiful virtual summaries of their learning. Students who loved watercolor used paper and paintbrushes to create beautiful, tangible summaries of their learning. Students passionate about bad math jokes used their humiliating amount of pun practice to create beautiful, transcendent summaries of their learning. I, as you may be able to presume after making it this far in the book, aligned with the latter, and thus, "Trigmund Freud" was born. By the time I created him in my junior year, I was no stranger to the summary of learning assignment. I had created "The Giving GeomeTREE" and "My System's Keeper" (following a tearjerker theme) before reaching the mindset capable of conceiving Trigmund Freud (less of a tearjerker, more of a jerk).

Humorous as it may seem, the process of creating Trigmund brought me great joy, resulting in my ongoing recollection of the different aspects of trigonometry. The process of learning to maintain knowledge is beautiful, as is the final exhibition of that knowledge. Math, being the subject-next-door, is beautiful with and without glasses on.

INVITATION TO
REFLECT

- At what point in your math learning did you experience the beauty of math most deeply?

- What work of your students is beautiful enough to hang on the wall for visitors at the school to see? How can that beautiful work become a launchpad to even more beautiful work in the coming years?

CHAPTER 8

DEAR MATH,
YOU ARE FUN

*"Dear Math, It was fun to see the numbers make
new numbers in new and cool ways."*

PARTWAY THROUGH CLASS on a Friday, a student called out, "Can we do a Kahoot? Please!?" Immediately, at least eight other voices echoed, "Yeah!" It should be noted that eight voices can *feel* like everyone sometimes.

The previous week, we did a Kahoot that one of my colleagues had made on general trivia about the school. The game was fun, and we learned new details about the school and each other through the experience, but I hadn't anticipated their assumption that this activity would become a part of our regular Friday routine. Nevertheless, I decided to end our activity early that day and try a Kahoot.

While the students were collaborating, I quickly searched for a Kahoot on the unit of study we were on in my eleventh grade class. I found a premade activity called "Transformations of Parent Functions" that seemed at least mildly relevant, and I clicked the "play" button. I noted that 24.6 thousand people had played this game, and I imagined all the fun in all those classrooms, with groups of students showing their understanding of transformations of functions.

When we started the game, I had the music muted. "Play the music!" some students exclaimed. I looked at them quizzically. The music? Is that an integral part of Kahoot? Apparently.

The students eagerly joined the game on their phones, and in just one minute, about 85 percent of the students were in the game. They had named themselves with rather unusual and silly names, so I wasn't altogether certain about who was and wasn't in, but we started anyway.

The questions scrolled on by. "The Parent Function $y = x^2$ is transformed to $y = x^2 + 5$. What is happening to the graph?"

The answers streamed in while the music blazed on. The cheers erupted with each answer popping onto the screen.

"D! Shift up five units!" We celebrated, we cheered, and we watched the rankings shift.

They seemed to be having fun. The students were smiling, they were answering math questions, and they were, hopefully, learning. The voices of the loudest and most competitive students rang loudly. I noticed at least five voices were not present in the joy or mathematics. Those students were sitting and playing and sometimes smiling, but for the most part, I felt uncertain about their experience. Did they want to win? Did they believe they could win? Was this game increasing their feelings of being mathematical? Was it expanding our mathematical community?

The game ended predictably, with a young man who stands at the intersection of gamer and "I do math in my spare time" as the winner.

Second place was another boy, who said, "I just guessed on every answer."

Third place was a girl who sat below everyone's radar the whole time, carefully reasoning on each question and answering with precision and enough speed to stay in the rankings. The screen showed their "names" bouncing up and down on a podium, and the winning student smiled.

The class period ended, and the students streamed out. As I stood alone in the room, I couldn't help but carry a healthy dose of skepticism about the "fun" that had just transpired in the room.

STUDIES ABOUT FUN

What do you think of when you hear the word "fun"? Does it involve facial expressions? Being with certain people? A particular activity? Games? Math class?

Fun is defined as enjoyment, amusement, or lighthearted pleasure. In the student letters, the word fun arises with a fair amount of frequency with respect to math classes. We see the word arise in a few situations. For example: "In elementary school, you were fun because we got to play with puzzles." Other students speak to math being fun in elementary school because it "was all fun and games" or "made sense" or, more frequently, "was easy."

Another context around "fun" is related to understanding: "When I understand you, you're fun, even on tests" and "I liked solving a challenging problem."

These show different conceptions of how math can be fun. Travis Tae Oh, a psychology researcher, explains, "My research on fun shows that, although a myriad of activities—traveling to a new city, riding a roller coaster, meeting old friends, watching a movie, going to a concert, etc.—can be considered fun, the intensity of the fun experience rests on two psychological pillars of hedonic engagement and a sense of liberation. Having fun, in fact, is an experience of liberating engagement." The student language of "play" seems to fall in line with hedonic engagement, and the language around learning something new seems to fit the liberation piece. In fact, if I were going to recreate these axes for math class, I might use two variables that are slightly different: "opportunity to play" and "deeper understanding of something that didn't make sense before." Since we focused on ways students consolidate and transfer learning in the last chapter, let's dig a little more deeply into the play factor.

In an article titled "The benefits of play for adults" and published in Help Guide, the authors state, "While play is crucial for a child's development, it is also beneficial for people of all ages. Play can add joy to life, relieve stress, supercharge learning, and connect you to others and the world around you. Play can also make work more productive and pleasurable."

The student letters speak frequently to a "loss of play" as they got older. There is a distinct sense that they used to feel they were playing around with numbers and making sense of shapes, until a point where the fun and games stopped and it became "hard and confusing." This is a problem in more ways than one.

Why does the play stop? What takes its place? Are Kahoot games the panacea for lack of fun in all math classes?

I would like to start answering these questions by outlining portions of "fun" that I am skeptical about, namely, the gamification and gimmicks of math class. While these activities and routines may have a place and can possibly provide a point of entry for students who are struggling, my ultimate recommendation is to sprinkle in these cheap thrills sparingly and intentionally.

"Dear Math, In elementary school, I thought that you were hard. I thought that you were so hard that I would never ever get you and how you are important in everyday life. But as I got older and older, I gradually learned that you are really fun, and I found ways to make cool art using you, like there are these cube art things that have math equations in each cube that equal a certain number, and that number would be a color, thus getting this cool image. That's what I mean when I say you are fun. When I went on to middle school, sure, it was hard at the time and so is life, but you get through then just like I did with your math in middle school. For high school, I don't know what changes I will face, but I know this: I will get through them." -Romero

THE GAMIFICATION OF MATH CLASS

Schools, and math classes in particular, have recently utilized gaming strategies. If we believe that math follows a linear sequence with clear learnings at each step along the way, then it would make sense that the learning could be video game-esque. When you pass a level, you earn a magic star, piece of armor, or big bouncing letters to get that quick dose of dopamine before moving on to the next level. While video games can be fun, and many people (students not the least of these) would cite "fun" as the way they describe these games, I am somewhat skeptical that this type of fun is what will create deeper thinkers and learners in our math classes.

Why am I skeptical? First, these games can often turn math class into a race, both through the content (skimming along the surface) and against the others in the class. As we already know from the last two chapters, the facets of race and competition serve to marginalize some of our learners and paint an improper picture of the true work of being mathematical.

Naturally, you might ask, "Should we celebrate students who like the competition and are quick to find the right answers in their head?" I am not in the business of tearing students down, so I celebrate this way of being mathematical alongside the rest of them. However, I am careful not to elevate speed above any other qualities.

My son loves numbers and spends lots of time thinking about numbers and patterns in his head. As a result, he computes arithmetic, creates proportions, and reasons about quantitative situations fairly quickly. Sometimes people ogle at him as some sort of mathematical savant. My goal is to celebrate his thinking and then ask if he could think about the problem in a new way. When we debrief his school day, I ask about his classmates' strategies and math ideas that he found intriguing. This doesn't downplay or deter from his mathematical growth; it deepens and strengthens it and centers on the idea that people who aren't as quick with number operations are still deeply mathematical, and his work is to celebrate that.

GIMMICKS OF MATH CLASS

We are living in a TikTok generation where content is quickly thrown at us in fifteen-to-twenty-second bursts of visuals, words, faces, and music. Recently, my algorithm has been feeding me videos from a math tutor claiming to show tricks your math teacher never taught you. The videos offer tips such as: when you're comparing two fractions, multiply on the diagonal and compare the product, and the side with the larger product is the larger fraction. The video host smiles as if he has given you a gift from the mathematical gods that your teacher has been holding out on you forever. I've taken to commenting on these videos, "Tricks and gimmicks are not math!" Math is a creative and sense-making subject, and when we reduce it to a few tricks to keep in your pocket, students are likely to misuse the tricks or develop a totally isolated understanding of topics and little contextualized understanding of where they fit into the world of mathematics. In the book *Nix the Tricks* by Tina Cardone, we read that the tricks of math undermine reasoning and sense-making. While having a trick up your sleeve can feel fun in the moment, it does not ultimately build stronger mathematicians.

Of course, fun and play are important in life and in math class, but some types of fun are cloaked in gamification and gimmicks that undermine community and the true spirit of mathematics. So how might we spark fun and play in our math classes with depth and authenticity?

Jayne was a pre-service teacher in my Math Methods class. In her final presentation of the year, she stated, "I want students to feel okay being silly in my class." Jayne is a thoughtful and reflective practitioner, and her quest for "silliness" stood out to me. I sensed that the silliness she was striving for was neither cloaked in gamification nor the gimmicks that sometimes plague our math classes. I sensed that she strove for the type of silliness where students were free, excited, curious, interested, and entertained. After spending time in her classroom, I know this is precisely the type of environment she created for her students. I've found a few types of play that foster fun in a classroom and promote deeper learning.

MATH MEMES

Math instructor Howie Hua is one of my favorite people to follow on Twitter. He creates memes that are grounded in mathematics and are sure to get you laughing out loud. His followers often join in the fun and contribute alternate language for the memes or share another meme that they made. This gave me the idea to bring fun and joy to math class, so I gave it a try in lieu of creating summaries of learning for their portfolio.

I asked them to create a meme or a joke that represented their learning during our previous unit on logarithms (logs). We looked at memes and discussed why they were funny, how they were constructed, and how they showed a deep understanding of the topic. We co-constructed the following set of norms for creating a successful meme:

- relates to logs in a mathematical way
- makes you laugh a little bit at least (a chuckle)
- artistic in some way (looks nice/professional-ish)
- post it to a math meme account and someone likes it

The students performed this assignment beautifully and had lots of laughs along the way. As they critiqued the memes with each other, they had to assess why the construction of the meme made sense or was mathematically provocative. The conversations in this critique session were deeply grounded in mathematical reasoning and sense-making in a way that I might not have come to without the memes. Can memes be used in a "gimmick" way? Sure. But can they be used to bring joy and seek deeper learning as well? That's a definite yes.

MATH JOKES

I've worked with many teachers throughout the years who loved to use math jokes to bring a lighthearted spirit to their class. Once a week, they might change out the joke for a new one to keep the kids laughing. I have used this practice to foster a deeper understanding of a topic and then, as with the memes, challenged the students to come up with their own jokes for our current unit of study. Rather than having jokes about random "tan-gents" in the middle of a quadratic functions unit, I find a joke about quadratics for each week and share it. It is important to have just one joke a week because the joke has time to set in, and students who didn't quite understand the joke the first day can let it marinate in their brains and be included in the laughing. I have found that when a lot of jokes are used in succession, especially if they are about a large swath of topics, it serves as less of a community-building activity and more of a separator of the haves and have-nots regarding understanding the jokes. It then begins to bleed into the gimmicks category of practices, and I intend to avoid this entirely. In addition, students can be encouraged to write their own math jokes and bring them to class for extra credit or just for fun and enjoyment.

PLAY WITH YOUR MATH

Joey Kelly and Xi "CiCi" Yu are the curators of a beautiful site called "Play With Your Math." The introduction to their site invites students to see that the best reason to learn math is "because it's fun!" They state, "Playing with math can be as rewarding as playing sports, games, or instruments." The exercises on this site are open, provocative, and exploratory. When implemented in a way that centers student ideas and thinking, they're a lever for fun and intellectual empowerment. I often use these problems during the first week of school. One of my favorites is the "plus-minus" problem or the "4-4's." Every year, I have a bit of self-doubt that perhaps this group of students is the one that will not give a rip about these puzzles. Nevertheless, year after year, my doubt is squashed as students dive headfirst into these puzzles and come up with new ways of thinking and exploring that I hadn't seen before.

Giving students the opportunity to play with their math can shift their paradigm. My colleague Chris Nho used Play With Your Math problems for summer school a few years back. Each day, the students logged onto Zoom and engaged in an exploration of a problem. Often, one problem would last for an entire week, and at the end of the week, they would share their exploration and learning from the week, both in content and skills. This group of students was unsuccessful during the regular school year, but by using play in math, they were able to create new identities as mathematicians grounded in reasoning, problem-solving, and, most importantly, fun!

"Dear Math, I have always enjoyed challenging my brain while doing a new problem or learning a new skill or way to solve a problem. It was fun to see the numbers make new numbers in new and cool ways that would have made sense in second grade. In math class today, I feel that

153

math has gotten harder, and though it is still fun, I don't think it's going to get easier. Which is unfortunate, but it means I still get to learn more things, so it could be fun. In the future, I hope that math is still fun and I will want to learn more and do math instead of not wanting to do it." –Seventh grader

MATH PLAY TABLE

Most of the students' Dear Math letters that cite "fun" talk about elementary school because there was a sense of play, of not worrying about answers, and of just thinking and making sense of puzzles and manipulatives. There is absolutely no reason why middle and high school students would not benefit similarly from manipulatives. Math education leader Sara VanDerWerf writes about a play table in her classroom, a space where her high school-aged students can play around with mathematical items and create beautiful patterns in tangible ways. She has a couple of rules for this play table on her eponymous blog. I highly recommend reading her entire blog and also getting a math play table to help students continue to feel the playful fun and exploration they once experienced in elementary school, even through their middle and high school years.

Having fun in math class is integral to the brain development of our students, and it contributes to a classroom community founded in joy and a positive math identity. And we don't need to sacrifice sense-making, deep thinking, deeper learning, or the innate beauty and intelligence of all learners to achieve it. We also don't need to leverage our authority in a way that makes math about tricks and right answers as determined by the teacher. All we really need is a few routines grounded in our core values and in the needs of all students, and we will create more smiles and laughs each day.

"Dear Math, Sometimes you are very fun to do. For example, geometry is one of my favorites. I really like learning with shapes. Something that makes math much easier for me is using items that represent the quantities of math. My past experience with you has been a mix of fun and not fun, of stressful and non-stressful. You were easy in kindergarten and elementary, but once you go up into middle school, it gets way more difficult, or at least that's one thing I noticed. You can be stressful as well, if you are solving a problem that requires a lot of brain use, if you don't know the answer, and if you have tried many times but still can't get it right. When that happens, it makes some people just give up." —Jeffrey

FUN IN THE CLASSROOM

Hector was a student in my class who had a lot of mathematical brilliance and an uncanny ability to not take himself or the mathematics too seriously. In his letter, you can read the beautiful ways that he composes a story about what it means to be silly. He ended up reading this out loud to the whole class one day, and he had us all laughing hysterically. It was one of my favorite days.

"Dear Math, I'm not going to lie, I kind of like you. No romance. When I understand you, things are fantastic. I could be with you all day. $2x+10=20$? Fantastic! I really like spending time with you when things are like this. However, I don't understand you 80 percent of the time. Every year, you come up with some new, weird formula I have to use for twenty different scenarios, and it's very frustrating. I cannot stand you when you turn like that. I wish you understood how much I don't understand you.

I do need you, though, for taxes and milk. I wish I could leave you, yet I always find the need to get back with you. You're everywhere I go. So I guess I can't break up with you. I'm kind of stuck with you. Which leads me to my main question: What would you think of getting married to me?

I know, it sounds crazy. You're way older than me, see many other people (literally the whole world), I'm into women and don't even know what you are, and we don't get along super well. But think about it. We are kind of stuck together until I die. That could be decades! I'm not sure if you're religious or not, but I am. I'm pretty Catholic, and I can only marry Catholics unless I get permission from a Bishop, but that's a long process, and it wouldn't work super well if we had children (don't ask me how). I feel like raising them would be difficult with you, since we don't agree on many things like your order of operations or why you always see your x. But maybe this could work. We've been together for years now." —Hector

GIGI'S REFLECTION

Contrary to my thesis in the prior chapter, I do not believe math to be inherently fun. I do find the infinite complexity (fractal-like, one could say) of math to be beautiful and useful and powerful and all the other applicable adjectives, but it's also challenging. Don't get me wrong, challenging is great. Challenging pushes us to try harder. Challenging encourages us to become better. But challenging is not always fun. When people are deciding which movie to watch, they can't and won't pick *Inception* every single night; that would be far too much of a challenge on the brain (and would probably be a little bit too much Leonardo DiCaprio). Math can be both challenging and fun; the two are not mutually exclusive. This is not an easy goal to achieve; however, it requires teachers to think creatively when planning lessons. The perfect Venn diagram overlap of fun and challenging is the Everest and the bullseye of math education.

An instance in which I experienced this overlap was when we created logarithm memes in math class. Logarithms, while challenging, aren't highly germane to the real world; therefore, they lack a great deal of relatability and fun. Despite that, to this day in my college screenwriting classes, I find ways to reference logarithms (much to the joy of my fellow classmates). I attribute the bulk of this retained knowledge to the meme-creation process.

I credit the meme-creation process for a couple of reasons. Namely, the slight undercurrent of comical competition. With the ever-present anticipation of exhibition time, we all, in friendly spirits, of course, aimed to have the funniest meme. An integral element of creating the cleverest logarithm meme is a genuine understanding of logarithmic intricacies. The meme I ended up showcasing was one that included images of a basic Winnie the Pooh and a noticeably classy Winnie the Pooh with an overlay comparing Logex and lnx. That distinction between the two terms (rather, the lack thereof) remains in my memory because of the goal of creating a funny meme, the learning process I had to go through to achieve it, and the fun I had throughout the entire practice.

A similar embodiment of the humor-related push to comprehension came to fruition for me thanks to a book titled *Math Jokes for Mathy Folks*. I know the healthcare industry recognizes the applicability of humor, with laughter being "the best medicine," but it may be time for the education system to embrace it too. During one of my math classes with Sarah, we briefly read from the aforementioned book. In doing so, I laughed far too hard at far too many of these "mathy jokes." More importantly, I was unable to laugh at several jokes because I did not understand the math concepts behind their punchlines. This naturally prompted me to desperately seek understanding. I wanted to get the joke. I wanted to be in on the fun. Promptly after that class, I walked up to Sarah and asked her to explain the joke. She did so with a smile and a secondary joke playing on the original.

Be it from friendly competition, pre-exhibition pride, or misunderstood punchlines, the fun of math, although difficult to get to, can inspire the deepest of mathematical insights.

INVITATION TO
REFLECT

- What activities in your life foster the greatest sense of playful fun? Which of these elements can you bring into your math classroom?

- How can you balance gamified and gimmicky fun with a playfully exploratory and more communal feeling of fun?

CHAPTER 9

DEAR MATH,
YOU ARE USEFUL

*"Dear Math, I see you as a tool that is
needed to see, create, and learn."*

ASKED MY STUDENTS one day, "What do you think of when you hear the word 'modeling'?" After thirty seconds of thinking time, they did a popcorn share out. A few students shared comments such as:

- "A scale model? Like a small version of something you draw, and then you make a bigger one?"

- "Something that is a really good example of something?"

- "Oh yeah! Like a sample or an exemplar."

- "Something you build? Like Legos?"

The students were familiar with the term modeling and were able to connect it to concrete experiences in their lives. While the examples they shared were adjacent to where we were heading, none of them named anything about the potentially predictive use of modeling or the breadth to which modeling can be used in the world. Then, I made a claim to them that *any* question they could ask me could be explored using math modeling.

"Give me a question, and I will share a mathematical lens," I provoked.

"What should I get for lunch?" a student offered, bringing up the obvious question that was on everyone's mind during this third-period class.

"What are all of the variables you need to consider for this situation?" I asked.

Several students responded with variables such as how much money you have, what you like to eat, where you are allowed to go, what time lunch is, and what is most nutritious.

"Great! You are all underway with your modeling! Now, what is the most *important* variable?" I asked.

"The cost!" shouted one student.

"The taste!" shouted another student.

"Yep, you're right," I said. The students looked at each other, trying to figure out who I was referencing as right.

"*Both* of you are right! In a modeling situation, different people could attend to the same question and come up with different but equally correct solutions. It just depends on what variables are selected and what assumptions are made to simplify the model. You each selected a variable based on a set of assumptions."

We went on to unpack the assumptions that each student had made and discussed how every human decision could be framed in this type of modeling situation.

"*And*," I said, "every big company in the world has math modelers working for them, making predictions based on data and supporting strong decision-making, so it's an awesome field to go into."

At this point, I would like to say all the students joined in a collective cheer for the ways that modeling could impact their future life, but the clock clicked over to 12:25, time for lunch, so they all rushed out the door.

"Good luck with your lunchtime math modeling!" I cheered as they walked out the door. They smirked and scurried out to buy the tastiest or cheapest or closest options. I knew they would be modeling everywhere, and I looked forward to our work together on our upcoming project to explore math modeling and make good "food choices." Modeling, after all, is one of the most "useful" things a mathematician can do.

WHAT MAKES SOMETHING USEFUL?

For our purposes, "useful" refers to having a purpose in everyday situations for a variety of people, rather than speaking to a more theoretical need for something. Back in Chapter 3, we did a deep dive into what makes things feel necessary or unnecessary to students and the ways we can create an intellectual need for the work we do every day. This chapter focuses on experiences that highlight the usefulness of a current issue.

Let's return to the claim I made to my students that math modeling is the most useful thing a mathematician can do. In the Common Core Standards for Math Practice, math modeling stands out right in the middle. It reads:

> CCSS.MATH.PRACTICE.MP4 Model with mathematics.
> *Mathematically proficient students can apply the mathematics they know to solve problems arising in everyday life, society, and the workplace.*

It goes on to describe this as a critical portion of evidence for mathematically proficient students and how it might look for elementary, middle, and high school students. Connecting this to the student language, it uses math in a way that is helpful in solving everyday problems. This phrase from a student, Saul, stands out to me: "You see it in everyday life, and it can help you a ton."

So the Common Core authors cared about this and students care about this, but what might it look like in our math classrooms? Let me tell you a little about my journey of incorporating math modeling into my classroom, starting with my first experiences with math modeling as an adult mathematician.

"Dear Math, I see you as a tool that is needed to see, create, and learn. I'm going to need math in business classes in the future. You do a lot of good; you make it so we understand the world. You helped me find a plan for the future and solidify my career as an engineer. I hope we will work together forever." –Anjaneya

SHARED MATHEMATICAL MODELING

My first exposure to math modeling came when the Common Core came out, and I was working with my then-colleague Bryan Meyer to merge the new math practice standards with the project-based learning nature of the school where I work. We ran a workshop and hosted a group of teachers to collectively look at the idea of math modeling as a foundational aspect of project-based learning. Our participants created models to claim that Daylight Savings Time should or shouldn't persist. After leading this workshop and having the opportunity to think through modeling scenarios with my colleagues, I was hooked. I searched for more workshops and materials on math modeling and had the opportunity to attend an entire symposium on it at the Mathematical Sciences Research Institute, where I participated in a modeling task by Dr. Julia Aguirre about the Flint Water Crisis. I went into the next year thinking about all the ways the modeling cycle could be applied to any project or question.

"Dear Math, Sometimes you prove yourself very useful in tough situations." –Jimmy

My rollout plan with students was to do a shared modeling experience the whole way through as part of our rational functions unit. For our shared experience, we completed the Flint Water Crisis task together. As I worked on it at the workshop, I realized there were many possible mathematical goals for this task, and one was the creation and analysis of different equations to model a given scenario. I even noticed that a rational function could be composed if the variables were compared in a certain way. Just as Dr. Aguirre had done for me, I launched with an image from Flint and asked, "What do you know about Flint, Michigan?" The students made some brilliant connections between things they had heard in the news, what they saw in the image, and their life experiences. Then, they generated a whole set of questions about the image and the situation.

Here is a table outlining how the process went with the class:

Step 1: Define the problem	What problem are we trying to solve? What question are we trying to answer?	Is the donation by the Nestle company enough water to provide for the folks of Flint during this crisis?
Step 2: Define the variables	What factors are important to consider? What can we measure?	The group came up with the following list: • daily water needs • water costs • current population • changing population • truck loads • water shelf life • crisis duration • what is a calendar year? • bottle volume • number of adults • level of activity

Some groups decided to look directly at water needs of all ages of schoolchildren versus the water provided. To make this solvable, the age and water needs of the schoolchildren had to be averaged and the size of the bottles in the truckload had to be assumed.

Step 3: Make assumptions

What simplifications can we make?

How can we give ourselves a solvable problem?

The most frequently used variables were listed on the board and ascribed letters. Here is where we landed:
B = number of bottles
A = amount of water in each bottle
T = time (in days)
N = daily needs
K = number of kids

Step 4: Get a solution

Would a graph or other visual schematic help provide insight?

Can I hold some values constant and allow others to vary?

Different groups had various ways to make sense of the scenario. I selected and sequenced a few methods to share. Also, since I knew this would lead to an exploration in rational functions, I talked with the group that had an equation expressed as a ratio, knowing I would come to their solution for our analysis and model assessment the following day. It was critical for *my* mathematical goals that we move toward the relationships in rational equations. Together, we discussed which of the variables were changing and which would need to be fixed, given the assumptions. I also included this prompt in the handout the next day.

| **Step 5:** Analyze and model assessment | Does my answer make sense? What are the strengths and weaknesses of my model? What would I do for the next steps? | Each group created a poster to show the steps of the modeling cycle and their solution to the problem statement, backed by evidence. They also shared next steps for investigation, contributing to the idea that math modeling (and math as a whole) is never complete; rather, it is a continual process of revision and refinement of ideas while expanding the number of variables we were considering. |

Graph the inequality $y \leq \frac{6.5 \text{ million}}{x}$, y is the number of people provided for, and x is _____.

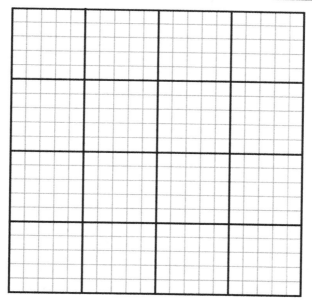

Describe the shape of the graph and what the graph tells you about the problem.

Image 9.1: An example of a classroom modeling sequence.

INDIVIDUALIZED MATHEMATICAL MODELING

For the students with the shared modeling experience, vocabulary like "problem statement," "variables," "assumptions," "building solutions," and "assessing solutions" became a regular part of our language. Each time it came up, even if it was in a relatively uninteresting area such as an SAT problem, I would reference our modeling cycle.

"Dear Math, At the very least, I have come to respect you. Your vastness and grandeur, your infinitesimally minuscule details, but most of all, the way everything just makes sense with you. There's some itch, something deep within the conscious that gets scratched when you start putting the pieces of you together. How the building blocks stack so neatly, how it seems each topic is like a three-dimensional puzzle piece, and each other topic begins to lock and twist to form a shape incomprehensible to any one person. You are order, the only order that we can comprehend. All occurrences, all happenings, all mysteries, are plain and simple with your mathematical looking glass. Our fates are prewritten in the universe, where yours is the foundation." –Trevor

As the semester started to wind down and we were getting ready for exhibition, I brought the modeling cycle back to the forefront of their minds. The students were researching food choices in their humanities class and completing some reading and writing about the social, political, and personal implications of the choices we make. Math modeling was a perfect way to help them with this. We carefully followed the steps of the modeling cycle for their food choices projects. This time, however, they were all creating individual models, so our critique and flow were more grounded in the cycle itself rather than in the shared mathematical goals of the task. In fact, the most challenging part was that some students' models were graphs of systems of equations, while others were editable spreadsheets, and others were scatterplots with trendlines from data to help make predictions. Despite the divergence in this activity, the students created beautiful posters that hung on the wall at exhibition and engaged visitors in taking a critical and mathematical lens to their food choices. Image 9.2 shows a student poster for this project.

MICROWAVED POPCORN

Is it better for me to buy microwave popcorn or buy a Hot Air Popper Presto?

DEFINE THE PROBLEM

Microwave popcorn has chemicals inside the packaging that cause many health hazards when melted.

PRESTO

VARIABLES

- Nutrition: Is this good for me?
- Money: Cost of popcorn
- Environmental: Deforestation
- Time: How long does it take to microwave vs. air pop?

ASSUMPTIONS

- I assume people have a microwave
- I assume I save money by buying microwave popcorn
- I assume microwave popcorn is somewhat healthy for you

amount of diacetyl

$y=0.0004x$

$y=0$

ounces

cost

$y=0.03x+17$

$y=.12x$

ounces

KEY

= microwave popcorn

= hot air popper presto

SOLUTION

A solution that microwave popcorn companies can do is to take out the diacetyl within the packaging. Diacetyl is the chemical formula that gives that buttery flavor, but the effects are bad. So if they simply take that chemical out of the packaging, it would make a difference.

DRAWBACKS

- Diacetyl is so unhealthy for people
- Microwave popcorn costs more money in the long run

BENEFITS

- The Hot Air Popper Presto is a lot healthier for you
- Has no added chemicals inside

Image 9.2: Leah Sample's popcorn model poster for exhibition.

To conclude with the words of Dr. Ricardo Cortez, professor of mathematics at Tulane University in New Orleans, "I don't teach models; I teach modeling. I like that the models come from the students' ideas, but at the same time, it's difficult to know the mathematics that is going to come up because it's always different. So, my syllabus is put together at the end of the course." When mathematicians, math modelers to be specific, set out to solve a problem, it is unknown where the learning will take them. This stands in stark contrast to the answer-getting race through the linear curricula that many students in this country are accustomed to. If we are to give students the opportunity to experience the wholly useful nature of mathematics, modeling must become a central feature of our math classes. We must be okay with not knowing exactly where things are going and with being ready to tread with students into the unknown of a complex question that they (and possibly all of us) don't know the answer to yet.

For a sample modeling handout, see Appendix F.

SOCIAL JUSTICE MATHEMATICAL MODELING

In the talk I referenced earlier from Dr. Julia Aguirre, we were given an alternative to the traditional modeling cycle to add in an additional criterion of teaching for social justice. The initial portion of the modeling cycle isn't just "Define the problem." Rather, the initial phase is expanded to include the naming of a broad social issue and the building of civic awareness and a sense of action. These parts contribute in a cyclical way to the problem statement or "specific situation" as described in this iteration of the modeling cycle.

In this way, math modeling is fertile ground for creating a more just and equitable world, the world we hope to live in and hope our students will create for the future. As a result, we must heed the call to dive into modeling situations and explore not only things like popcorn and oat milk but also topics like racial profiling, climate change, border issues, tax laws, and abortion. We must integrate these concepts with students' work because the future of our country depends on their ability to critically analyze the world around them and use math to forge a path forward. Aguirre explains the importance of personally engaging in this work for our students and for us as math educators.

She states, "The dual focus of math modeling and social justice can help address current challenges in a culturally responsive approach to mathematics teacher education in synergistic ways. We have shown how a carefully crafted task rooted in social justice and a community-based topic can lead to challenging work on data analysis, making reasonable assumptions, creating mathematical models, and validating them. In turn, the teacher-generated models lead to new questions and discussions related to social justice that can also be analyzed mathematically, thus creating a fertile professional learning space to advance our mathematical knowledge, develop teacher awareness and critical consciousness, and create a path forward to normalize and humanize mathematics and its connection to the world."

As math teams, how do we dive into the modeling cycle together to analyze the ways that our system is providing an equitable and liberatory experience for all students?

Math modeling is useful on its own, but math modeling combined with social justice can be a powerful force for good.

"Dear Math, You are awesome and have done so much for humankind. Without you, we would never have built all of our cities and homes. In elementary school, I wasn't the biggest fan of math, and the way my teacher explained it was confusing. My favorite subject in math is fractions because you see it almost every day, and it can help you a lot. I used to think I was bad at math because I thought I didn't do so well in my math class in third grade. I struggled with time and thought. This made me feel bad at math. But now I know I wasn't bad; I just didn't understand it. Now that I understand, it is so easy. My teachers explain everything clearly, and math isn't that hard for me anymore. Thank you for all you have done for us, math. You are a great thing, and without you, I don't know what we would do." —Saul, seventh grade

SOCIAL JUSTICE IN THE CLASSROOM

During my final semester teaching seniors, my teaching partner and I ended up designing mini-project-based learning units, and the students could opt in to the project group that most intrigued them. I ended up designing a three-week intensive course entitled "Math for Social Justice." A couple dozen students joined, and we met weekly to explore math and data analysis methods that can unearth issues of social injustice and help advocate for justice in a data-driven way. I structured the class around three issues from the book *High School Mathematics Lessons That Explore, Understand, and Respond to Social Injustice*. We diverged from the lessons using the modeling cycle to help us consider other variables we care about and how we might create models to seek justice.

One week, students looked at various US-Mexico border issues and did some independent research on impactful variables. Then they presented the variables they had explored and watched each other's presentations. The unit culminated with the groups explaining their research and the knowledge they gained from other groups, followed by advocating for the next steps in the modeling process.

The response ranges were vast, and while some explicitly named mathematical tools they would use to explore this more deeply (such as F tests and testing for correlation), other students advocated the government's release of more data so we can actually *do something* about the situation.

Specific quotes from students after these three weeks were:

- I noticed how much data can look completely different based on who is writing about it.

- Immigration policies in the US are related to who is president. I can contribute to solving the growing border crisis with donations.

- I'm confused about what some of the data means. Am I supposed to think that because fewer people are apprehended here, they are better taken care of? It's difficult because a lot of things are happening to these people behind closed doors. There are problems that are terrible, and I wish the government would do its job. I don't know how I am supposed to help except by staying vigilant, caring, and not letting this slide under the rug. Keep fighting for it.

- My main takeaway is about the total quantity of apprehensions.

- Something I learned about was Quentin's F value.

- Awareness helps, but also, if you have time and money, please donate.

- I went into helping families impacted by the immigration policy. I learned that kids are separated from their families, and a lot of them were separated in 2018–19. Three hundred of those kids were under the age of five.

- My takeaway was the scale of the media bias.

- Something I learned from someone else was just how many apprehensions take place each year. It was a much bigger number than I had anticipated. I think the most important way to help is to get educated about the issues, and then when you can elect officials, do it.

- One surprising takeaway was that there are so many apprehensions each year, and it depends on the political climate that year. You can see how much political leadership impacts immigration. I learned how many resources we have already available to asylum seekers and how much more they need here and in other places.

- The border problem is a huge problem, but we can look at it through a bunch of different lenses and learn more through the data and the needs and gain a bigger understanding of the problem. From my research, I learned that each apprehension costs $10,800. From another group, I learned that there is a lack of information, and the data isn't readily available. Sailor and I ran a successful food drive to donate food to migrants in need.

- I learned from another presentation that they don't get the supplies they need.

- I noticed how biased and skewed graphs can be, and depending on how they show their data, the authors can tell whatever story they want.

- The government doesn't publish a lot of data on this, and one thing we can do is advocate for the government to publish more data so we can make better sense of the situation.

GIGI'S REFLECTION

One fact that even math's most adamant critics cannot argue against is its elemental usefulness. The earliest applications of arithmetic, in many ways, helped shape our society into its contemporary form. Euclid's postulates, axioms, and common notions make up the foundation for geometry and deductive reasoning, catalyzing much of civilization's modern architecture. Surely, Euclid would consider his work to be rather elemental. Even when not discussing the father of geometry, it is important to treat math with respect because it wondrously exists at every turn. At the root of our universe, our communities, even ourselves—made up of billions of strands of patterned DNA—is math. Even stepping away from that pedestal where math's influence rightfully sits, we can discuss the prevalence of math in everyday life.

One of the most classic pieces of dialogue that likely plagues every math classroom across the world is, "Okay, but when am I *actually* going to use this in my life?" Even when uttered with the appropriate teenage snark, I *actually* deem this a fair question. Why would anyone care about learning something of no use to them? I would roll my eyes, too, at the thought of learning something simply because it's in the curriculum. Wouldn't you? It is from a quick, pertinent response to this question of usefulness that true engagement is born. The antitheses of this engagement are instruction and assessment styles that exemplify the arbitrary application of math, most notably, standardized tests.

One of project-based learning's most highly developed purposes is to find and teach applicability within math. The common question of usefulness is notorious for sending shivers down the spines of math teachers, yet my teachers (usually Sarah) were capable of answering coolly and correctly. Better yet is when the premise of the project outlined at the beginning of the class is so relevant that it does not provoke the question at all.

One such instance occurred during a project on taxes. It's funny how a group of adults groans at the mention of taxes, and a group of teenagers almost cheer at it when it's voiced within a classroom setting. Taxes exude applicability and usefulness. Hearing adults groaning about taxes is the reason we feel inclined to do everything in our power to learn about it: we want to avoid feeling the same way. In this project, we defined and discussed the likes of income tax, regressive and progressive tax, and student debt (as a current college student, I can adamantly say that this last one is especially useful). It was equally fascinating and inspiring to see a room full of sixteen-year-olds totally engaged in the details of taxation mathematics.

Yet another couple of projects that rivaled the usefulness question focused on math's reach into the realm of social justice. One project concentrated on the characterization and subsequent location of food deserts, using our newly gleaned knowledge of radians. Another challenged us to identify the mathematical variables behind a chosen food or nutrition-related product and make a model detailing them. I—the stereotypical Southern Californian, dairy-hating elitist that I am—chose to focus on the production of almond milk. I learned about the process of defining variables and making assumptions when modeling, as well as the effects of almond milk manufacturing on the environment and economy.

Not once while participating in these projects did I, nor any of my classmates, ask why the math lesson was useful. At some point when you were a student, *you* asked the question. Maybe just to yourself, maybe to your friends, perhaps to your teacher. If you were learning about the likes of architecture, music, taxes, or social justice, would you still have questioned it? Or would you have already known the answer?

INVITATION TO
REFLECT

- What current events could be explored using the math modeling cycle?

- What are your students learning about in their other classes, and how can a math model help them make sense of those ideas more deeply?

·

CHAPTER 10

DEAR MATH, YOU ARE POWERFUL

"Dear Math, You're my only logic in a sea of creativity and subjectivity."

THE 2020 ELECTION was upon us, and our math class was embarking on a journey to understand our election system more deeply and mathematically. I asked a student to volunteer to read the essential questions for our project out loud, and a student read the following from our shared project description:

ESSENTIAL QUESTIONS:

- How does our voting process work?
- Which voting methods are best?
- How might mathematics help end gerrymandering?
- Can data be presented in an unbiased manner?
- Can data predict the election outcome?

Building from my frequent claim around modeling—that math can help us answer any question better—I leaned into the notion that it would help us as a class better understand the election, even going so far as to claim that data can or cannot help us predict things like elections. I had heard that an esteemed colleague, Mele Sato, had done an elections project before, so I reached out to learn more and see how I might use her resources as a backbone to the project with my students.

At our school, teachers are centered as project designers. However, I had learned from my fourteen years there that inventing totally new projects didn't always lead to the best work. Sometimes, taking an existing project and modifying it slightly to meet my students' passions, talents, and curiosities—and my interest—would result in better quality work and greater collaboration.

PROJECT PHASES / LEARNING GOALS / PRODUCTS

	Learning topics	What will we create?
Phase 1 (1.5 weeks)	**Voting Methods** ☐ Electoral College ☐ Plurality (most votes wins) ☐ Borda Count (ranking with points) ☐ Approval (one point to each candidate you approve of) ☐ Score (approval from 0 to 2, 3, etc.) ☐ Condorcet (A over B, B over C, etc.)	**Voting System Proposal:** • Description of proposed voting system in words and with an example • What voting criteria are met under the new system? • New method should meet at least 3 of the voting criteria • Give an example of how your voting system meets at least 1 of the criteria • How is your proposed voting system more fair? • Pros/cons of new voting system

| Phase 2 (1.5 weeks) | **Gerrymandering**
• Compactness
• Proportionality
• Efficiency gap | **Redistricting Proposal:**
1. Current congressional district map and their respective measures of compactness
2. Proposed congressional district map with at least two measures of compactness completed
  a. Why did you choose those measures of compactness?
3. Explanation for WHY you modified your district in the way that you did
4. How are your proposed districts more fair?
  a. Pros/cons of new districts |
| Phase 3 (2 weeks) | **Bias in Data**
• The "stories" that data tells us
• The ways that data is used to manipulate voters
• How does (unbiased) data help persuade voters (yard signs) | **"Need to Know" Yard Signs:**
• Title/topic
• Authors names
• Key facts ("need to knows")
• Benefits and issues
• Example of topic in action |

Image 10.1

Throughout the project, we worked through various mathematical elements of the election using benchmarks. Students created voting system proposals and redistricting proposals. A variety of guest speakers shared their knowledge of predictive modeling, and we attended workshops put on by Five Thirty Eight to better understand the ways that mathematics was already being used to make sense of this complicated election cycle. The class also made predictions about the upcoming election, particularly considering the mispredictions of the election four years prior.

While one of the final products would be yard signs, the remote nature of this semester made this challenging. The project took longer than expected, and the signs were not going to be ready by the election. Instead, we pivoted, and students worked in groups to propose a new system of voting for associated student body elections. The proposals were compiled and then presented to current ASB members and school administration.

"Dear Math, Even when you became increasingly more difficult, I found the challenge intoxicating. I spent hours researching you, planning on becoming a professional studying you in later years." –Penelope

The students also voted among themselves for the proposal they thought was most effective. They justified their choices based on the modeling they had done and, as seniors, realized they had left a lasting mark on the school's fair voting system. While the project work had an audience and benchmarks along the way, the power of the project lay in the students gaining a lens for their power—to make sense of a system that might be broken in one way or another and then to advocate for a new way for that system to run. Students realized that competitions exist for proposed redistricting for states and new fairness measures, and they learned that they could participate in changing the voting process.

While the students had done election-related projects before, I heard a reflection that this was the first time the thesis wasn't "get out and vote because it's important to make your voice heard." In this project, the students got to interrogate the voting system as it was and imagine an even better system. This is the ultimate powerful work because it touches on the usefulness of mathematics and the identity, agency, and growth aspects of a project by providing an opportunity to become a better citizen through the experience.

"Dear Math, When I was younger, I always thought math was difficult. I thought I would never master the skills of math. But now that I've seen and learned more, I really like doing it. I used to think to myself, "I don't know why I don't understand this; I hate this!" But now I don't see a problem with doing it; I've become way better than before. ... I knew that it would be difficult, but I also knew that if I try and try, I would eventually master it. ... It's always hard for the first time learning a new subject, but I understand it with a little bit of effort and work. One time in sixth grade, I did very well with my partner. I was so happy doing so well." —Junior, seventh grade

POWER DEFINITIONS IN MATH

Power in various contexts can involve strength, control, or influence. I have already shown in this book that math holds power in society (possibly disproportionately) to make people feel smart and elevated or lacking and intimidated. I have also written of the ways that teachers hold power in a classroom and the importance of considering how we wield that power to uplift student ideas and regularly celebrate students' mathematical brilliance. The power in this chapter, however, is a feeling expressed by the students. While students themselves rarely use the term "powerful" in their letters, a certain category of student responses spoke to something big and life- or mind-altering, and the best classification we could think of for these ideas was "powerful." Let's unpack the student expressions of "power" with a bit of analysis.

"Dear Math, No other subject has captured my confidence like you have. People say that I'm gifted in math, but I would like it just as much if I had to struggle through everything. I love the feeling of learning a new concept and having trouble with it at first, because learning it in the end is so rewarding. I think that I hold onto you so much because at our school, there aren't any other subjects that we practice like you. You're my only logic in a sea of creativity and subjectivity." —Conner

First, there are quotes like Penelope's and Conner's, which use language like "You captured my confidence" and "The challenge was intoxicating." These speak to a feeling of math having power over a person and their emotions or of being truly "taken" by the study of it. This is adjacent to another idea we see in student writing, which is that math has the power to give students something they were missing. This is seen in quotes like Trevor's and another part of Conner's letter, where they say things like "You're my only logic in a sea of creativity and subjectivity" and "You scratch an itch in my unconscious."

I particularly like the language that speaks of the tremendous power that math can have for people to think and reason more deeply about something deep inside themselves. Math was originally a field of study explored by philosophers, and to this day, the "existence" of math or the "discovery" of math is a hotly debated topic. Some folks claim that math ideas are concrete and exist with or without humans to know them, and others argue that mathematical truths are based on human-made rules and cannot exist without humankind's thought. It's a wonderful rabbit hole to go down on an afternoon when you have some time on your hands (search "was math invented or discovered"). For the sake of this chapter, however, let's just say I appreciate the students who express language that dabbles in either of these spaces and voice the power of that feeling over them.

"Dear Math, Some of my friends think that math is the study of everything and the world wouldn't function without it. Some people tell me that it's not that important to learn math. I think math helped me try to understand and work through problems even if I was doing something other than math. I think it helped me be more educated and quick to answer problems and questions. I think I like to write and solve problems step by step and neatly. I'm glad I have that as my greatest mathematical strength. I think one of my greatest mathematical challenges is algebra and creating and solving equations. .... I can thank math for being so broad and surrounded in so many different studies." —Max

Next, we come to a collection of student quotes that speak of the power math has to help individuals and the world. The ideas felt to us like math has an extra element of "power" beyond just the utilitarian need. These quotes sound like "Everything makes sense with you," and "You are the greatest gift to humanity," and "I can thank math for being so broad."

Some students' letters attest to the power of math that allowed them to see themselves in a new light. These are counter stories to Chapter 5 (Dear Math, You Are Oppressive) because they document the ways that mathematics empowers students with new perspectives. These sound like "You help me understand and work through problems even if I'm not doing math anymore" and the growth mindset message, "I believed that if I worked hard, I would eventually figure it out."

Finally, one student statement has stuck with me for a long time. While not echoed by many other students, this writing empowered me as I read it and has provided helpful language to other adults I have shown it to. This student, Anjaneya, discussed math with an animalistic power.

"Dear Math, You're a beast. Only manageable when tamed, but once you are, you're loyal and helpful." —Anjaneya

In this description, I can't help but think that Anjaneya is speaking of his experiences with the immense power that mathematics has over him and has given to him.

In her blog on Creative Maths entitled "Mathematics is Powerful," Dr. Nic speaks to these same themes that have emerged in her conversations. She says that the elements of mathematical fun, play, and logic that many find purpose in are great, but they do not encompass everyone's experience with math. She claims that "powerful" is exactly the word to describe math, as it gives credence to all the ways of knowing and doing mathematics in service of self-growth and betterment and for world betterment. She states, "At a global level, mathematics saves lives, powers the internet, improves our daily lives, and will save the human race from destruction. At a local level, mathematics enables businesses to prosper, allows us to find our way, helps us make sensible food decisions, and allows us to create works of art."

People who are given opportunities to experience the power of math at the global level and local level can then help themselves from the individual identity level, as in the cases of middle school students Max and Junior. It seems from their quotes that doing math truly changed the way they thought about themselves, which is powerful, indeed!

In the last chapter, we talked about the process of mathematical modeling and the ways it can unleash the power of math in the real world. I would like to, by extension, use this chapter to speak about the powerful nature of project-based learning and the role that math plays in it. All the student quotes in this book are from students in the PBL schools where I work, and, while this does not prove causality, I have found fewer accounts of these types of powerful emotions in schools that do not leverage many of the principles of PBL.

While I have worked at a PBL school for 95 percent of my years as a classroom teacher, and therefore many of my beliefs have been formed around this method, I believe that what follows can be useful even for teachers employing problem-based learning and teachers just hoping to add depth to their student-centered lessons and activities. I am a strong believer that PBL isn't a stand-alone practice and that all the routines discussed in previous chapters are necessary elements of any equity-oriented, student-centered math classroom. In that way, I would say my entire book has been about project-based learning as I have come to know it in my context, but it is not true that the practices in previous chapters make up and apply only to PBL.

How do we know that project-based learning is an effective practice for supporting student achievement? For a long time, PBL schools like mine have relied on anecdotal evidence of authentic student work, high engagement, and deep relationships, along with concrete data about college enrollment and persistence. Just recently, however, Lucas Education Research conducted a randomized, controlled trial of thousands of students in diverse school systems in the US and learned that PBL schools significantly outperformed traditional curricula. In fact, academic performance was higher across all grade levels and socioeconomic subgroups. While I was heartened by this research, I am also aware that PBL can be used with varying levels of fidelity and authenticity. Here, I am sharing a few examples of how math has been approached in my PBL school and the ways that my math team and I have come to see math and PBL as inextricably linked in a powerful connection. (For more details, check out the book *Project Based Learning: Real Questions. Real Answers. How to Unpack PBL and Inquiry* by Ross Cooper and Erin Murphy.)

ESSENTIAL QUESTIONS

Essential questions is a learning approach about the "how" and "why" of lots of different situations, and it adds a powerful framework to math class and PBL. The questions should:

- guide the project from start to finish
- be illuminated more deeply as the project progresses
- be open-ended and provocative

The essential questions of the electioneering project spoke to all these attributes and named clear learning outcomes that I hoped students would achieve by the end of the project. This approach may incorporate questions that don't seem to be math-specific, and the questions can branch across disciplines and be answered through a variety of lenses during the project.

Sample project essential questions might include:

- Why do humans need to protect the earth, and how can we as sixth graders play a role in this?
- How do we create values using only one color and black and white?
- How can high school students help to address issues of poverty and hunger in both our local and global communities?
- How can an election candidate effectively persuade voters to elect them?
- How can a home be designed to have little impact on the environment?
- How have the simple inventions of the past helped to create the complex technology we use today?

Ideally, the essential questions are designed so students can routinely look back at them over the course of the project and see clear and measurable growth in their understanding. I have known teachers who create posters housing the essential questions in their classroom for the duration of the project and routinely (weekly or biweekly) have students write about their new understandings related to the questions. These check-ins might come in the form of benchmarks.

BENCHMARKS

Benchmarks are stopping points in the project, either when a portion is complete (if the project is designed in phases like the electioneering one described earlier), when a draft of the final product is done before the students head into a cycle of critique and revision, or whenever it is important to the learning process to check back on the essential questions and realign the project work to better meet the needs of the students and, potentially, the audience.

In projects where the benchmarks are emerging drafts of one final culminating product, I have leveraged benchmarks to have students create a new draft and reflect on the ways that our new learning helped us refine our version of the final project. In one of my projects, "Geometric Gardening: Succulent Style," the students used the tools of geometry to design hanging succulent gardens. The learning trajectory of the unit included the basics of geometry, geometric constructions, and rigid motions. The drafting process for the gardens went like this:

1. First draft of planter design using paper

2. Second draft of planter design with new partner for new ideas

3. Third draft on cardstock

4. Final draft using laser cut wood

The following text is from a Student Reflection documenting this process.

First draft:

"I collaborated very well with my partner; we were constantly planning ahead to the next steps we would take, and it really paid off. We completed the design challenge before our time ran out, and it was a functional hanging garden (it's just for plants so small you can't quite see them)."

Image 10.2: Geometric gardening project drafts by Kyna Airess.

Second draft:

"This photo represents how functional our second attempt at a hanging garden was (this time with my new partner, whom I would go through the rest of the project with). Its rings were too thin; we didn't think out properly even though we were being too ambitious and trying to do more than we were capable of. It didn't really hold *together*, and it didn't really support its own *weight*."

Final draft:

"After a few rounds of trying too hard and failing, we knew what we were aiming for: simplicity. Well, there still needed to be dodecagons; that was non-negotiable, but we made it simpler. You can also look at it from an angle that makes it all dark stained wood, all light stained wood, or mix it up and get a ring that's light and two that are dark or vice versa."

Benchmarks help students consolidate their learning and feel the power of mathematics through their confidence, and they also give students a moment to stop and breathe and reflect on the ways their growth in mathematical knowledge is leading to more beautiful and precise project work. Benchmarks help to redefine mathematics as continually learning and growing through mistakes as opposed to a more traditional structure of "learn, practice, test." I believe that it is through these types of PBL math classrooms that students can experience the full power of math for the world and themselves.

AUDIENCE

Another element of PBL that sets it apart from modeling as a stand-alone routine is the idea of an audience. In this case, the audience refers to the people the projects are serving. Many traditional classrooms have students completing work for the teacher and possibly for themselves to get good grades and head to the next step on their journey. Projects are specifically designed to create work for an audience beyond the teacher. Ron Berger of EL schools has dubbed this idea "hierarchy of audience." In his book *Leaders of Their Own Learning*, Berger outlines the different ways of thinking about the audience in a project-based classroom. He also says that as the audience becomes increasingly authentic (i.e., in service of the world in some way), motivation and engagement go up.

The audience should be considered at the outset of the project, with students asking questions like "For whom are we creating this work?" or "For whom are we doing this project?" Add questions like "What does our audience need, and how can we create something to authentically meet that need?"

For many projects, the audience is the students' parents and the school community members who are coming to the exhibition. For my geometric gardening project, it was also an opportunity to use mathematics to beautify the school and intrigue people when they came to the building. For other projects, students might share their work at a company or organization, and the presentation consists of real stakeholders in that company giving feedback on the final designs. One such instance we had was the Innocence project at my school. Students went to court to attempt to make a case for the freedom of an incarcerated person who had been in jail for a long time.

FINAL PRODUCTS AND EXHIBITION

The final product is what the students make for the project. It is intended to be shared with the audience and to represent all the learning throughout its creation process. It is "quality work," meaning it has gone through various critique and revision cycles in class and possibly with outside experts to reach its final draft state. We strive to provide the students with all they might need to share their best possible work on exhibition night.

Exhibitions can be big and public or small with a few people in the home. Exhibition night at our school is the biggest night of the year, and the school comes alive with student excitement and authentic work. We also routinely host exhibitions offsite if it proves to be an authentic connection and good opportunity.

EMPATHY EXAMPLE

One ninth grade project was called eMpATHy (because you can't spell empathy without MATH!). Image 10.3 shows eMpATHy posters that analyze a graph each student created from the classroom data set. For exhibition, they describe the shape of the graph, measures of center and spread, and how it tells a story of the class as a whole. Then each student shares an eMpATHy moment, connecting themselves to the graph and measures.

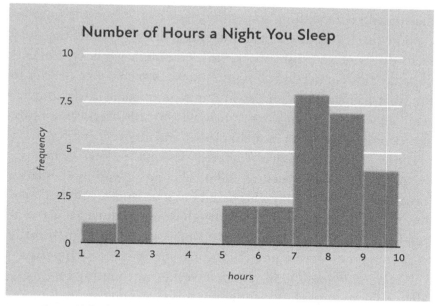

Image 10.3: This is a portion of an eMpATHy work sample. After the graph, the student added a narrative for analysis and an eMpATHy moment.

As alternatives to exhibitions, students can share their designs with a family member and then share out the family member's response to reflection questions. Also, students can share their designs with each other in a Zoom call. Students should also "exhibit" their work to each other along the way as they increasingly find connections to one another through the data.

SCHEDULED PROJECT WORK

I alluded to this at the beginning of this section, and now I'm looping back to the idea that true project-based learning does not happen only once a year following a unit of study where the students "take a break" from their learning to make something. Rather, in true project-based learning, student learning and product creation are happening simultaneously through thoughtful design and implementation. It is a misconception that while doing a project, the students are building something each day. It is also a misconception that in a project, the students don't do lessons that may seem disconnected from the project (although we hope students do make this connection along the way). As such, I like to refer to PBL that happens at all times in our classrooms as truly student-centered, equity-driven work. It happens in daily, weekly, monthly, and semesterly projects in math classes and all other subjects.

Daily: Regular student-centered lessons follow a launch-explore-summarize structure and a coherent flow of mathematics learning. Student thinking is shared out loud and "exhibited" regularly on the board to draw connections, and student collaboration is central to the classroom. Students learn how to functionally work in a group, leveraging the diverse strengths of each person.

Weekly/monthly: On a weekly, biweekly, or monthly basis, students refer back to the essential questions and make progress on their final products in iterative cycles as they complete benchmarks. Students critique and revise both their products and their mathematical understandings.

Semesterly: Students work on their final products, and the math work contributing to the beautiful results is shared at exhibitions.

At each of these levels of frequency, we ensure a sense of authenticity, collaboration, and exhibition.

TYPES OF PROJECTS

Depending on the structure of your school, your opportunities to do projects will vary. Even in my school, where we value PBL as our central framework, differing school structures have lent themselves to different elements of project-based learning work. Through all the shifts, we have persisted in our work and found ways to accomplish the central tenets of PBL, no matter our current schedule and team structure. Here are a few types of projects we have planned that illustrate the powerful combination of math and PBL:

Math projects as portfolios: While many PBL purists might disagree that this constitutes a true "project," I have sometimes, when working with new teachers, encouraged them to complete a first project that is structured as a clear unit of study in a high-quality curriculum, create a broad and content-embedded essential question to guide the work (such as: Why is it important to learn about quadratic functions?), and then, have students select and reflect on the work in a portfolio with a summary of learning. The students then take their portfolios home to share their learning with their parents or set their portfolios out at exhibition. While the tenets of PBL like authentic audience and exhibition might be weak in this type of "starter project," they can be helpful building blocks for teachers who still need to grow in their confidence and ability to facilitate a student-centered, discourse-based classroom and for those who are new to this type of content. In the absence of this pre-learning, I have seen teachers who design projects that are either way too basic in mathematics for their students or far more complex than the students are ready to explore (e.g., orbital mechanics in a sixth grade classroom).

Stand-alone math projects: In the absence of teaching a shared group of students with a "team" of cohorts and teaching partners from other disciplines, it may only be possible to design a stand-alone math project, meaning the learning goals are generally only tied to math. When designing for this scenario, consider the mathematical goals of an upcoming unit and how they might blossom into a beautiful project for students. Alternatively, you might come across a modeling scenario or beautiful display of work at a museum and think, "This could become a project for my students!" These may be shorter versions of the semester-long, interdisciplinary projects.

Interdisciplinary projects: Finally, a goal for educators working in PBL spaces is to design projects with an interdisciplinary team of teachers, where the teachers and students can collectively explore an essential question together with multiple lenses and subjects of expertise to provide new insights about the question. These projects are often longer, possibly even an entire semester, and the level of collaboration between the teachers may ebb and flow. In a world where there is an excessive separation between disciplines, this type of project design presents a beautiful future for education where students don't see all of their learning as broken into small, discipline-specific lessons, and they can blur the lines that exist between English, history, math, science, and art.

ENHANCING EXISTING UNITS

The use of curriculum can either enhance or deter the flow of a project. A high-quality curriculum can be a helpful tool for teachers as we continually improve our craft. It is tremendously difficult to be a full curriculum designer while also mastering the student-centered teaching style, always giving thorough feedback to students, and getting to know them well. It often leads to burnout. Instead, harness the power of PBL and "overlay" a project in an existing unit of study.

One strategy to do this is to take existing units in the curriculum and design essential questions, prioritize lessons, add benchmarks, suggest audiences and exhibitions, and create a few reflection questions.

In a world where schools have constraints and requirements, I hope you can see a variety of opportunities to bring elements of PBL to your daily, weekly, and monthly practice. It can certainly be a powerful experience for teachers and students.

POWER IN THE CLASSROOM

The day of our second public lesson study had come, and the students were gathered in the main commons of our building. The front of it was meant to mimic our classroom, and the back was where the observers were to sit. The students were noticeably nervous with all the visitors, and I can assure you that my heart was beating the fastest.

The lesson study team and I had spent weeks looking at student work, completing empathy interviews, designing the lesson, and then anticipating student responses to our prompt. The lesson was a riff on a lesson from *High School Mathematics Lessons to Explore, Understand, and Respond to Social Injustice*. The mathematical goal was to find bisectors that evenly divide a space between two points, and there was a heavy connection to the project we were currently working on called "Circling Food Deserts."

Mark, a boisterous and charismatic leader in the class, was sitting at the middle table in the front of the room with two other students. Mark carried a lot of social status, but he lacked academic status. In fact, I don't think he had ever felt like he was a mathematician and mostly used his charisma to direct people away from noticing his math insecurities. I liked to listen intently to Mark. I knew his mathematical brilliance was in there and just unrecognized.

The group with Mark worked diligently after the launch of the problem. They were trying to locate where two grocery stores should be placed on a grid to ensure that most people had access to the store. After completing this task, the group placed three stores and then were deeply pursuing the question, "How would a person know which grocery store was closest to them?" The group began by guessing and checking and asking each other questions about what might be best. Mark had drawn on his paper rather haphazardly, feigning mathematical engagement and looking curiously at his group members to avoid being pushed in his thinking whenever I walked by.

I knew this about Mark, so I didn't walk near with a watchful eye but rather listened to him carefully for a nugget I could use to highlight and share more broadly with the class. Fortunately, I didn't have to wait long! After about ten minutes of work time, I heard Mark exclaim to his group, "We should build a high-rise building right there!" Mark pointed at the newly found triangle circumcenter on the page—the point equidistant from all three grocery stores. I smiled.

Mark's connection to the meaning of that point was stunning and beautifully articulated. If a person is trying to find the spot on a grid that is the same distance from three grocery stores to ensure food access, they shouldn't just place a home there; they should maximize the number of people living in that spot by placing a high-rise building. Mark not only knew what the point was describing but had found his way to express his mathematical thinking and make a connection. And his connection had meaning because it was related to the powerful work of understanding food deserts, an ongoing project we had been working on in my class for the previous three weeks.

Students will often make unusual connections in mathematics, and Mark's decision to place a high-rise building on his paper was no less mathematical than the carefully crafted articulations using geometric language that some of his peers had made. Keeping my ears open for creative mathematical thinking is one of the greatest joys of my career. I employ carefully crafted projects that give students the opportunity to bring their full selves and creativity to the classroom each day. Through all of this, I have ignored my own conceptions of what it means to be mathematically correct and articulate and instead respond to and celebrate the students' own ways of being mathematical and creating important work for our project.

As the lesson ended and students shared their thinking, I asked Mark to share his brilliant idea about the high-rise apartment building with the triangle circumcenter to conclude the lesson for everyone. He stood and shared his thoughts. In his way, Mark helped everyone gain a deeper understanding of the lesson that day, and I gained a deeper understanding of the ways that project work can help students make sense of mathematical ideas. Thanks, Mark.

GIGI'S REFLECTION

It's time to delve more deeply into the power of math within the modern, complex, and problematic world. One of math's many beauties is that the entire reason it was created was to solve problems. Unfortunately, the problems of humanity are slightly more complicated than an expression requiring you to solve for x.

Luckily, "math" and "complicated" go together like peanut butter and jelly. As problems grow in convolution and breadth, math's ability to solve them grows in tandem. The true power of math lies in its application. Math as a concept is just that, conceptual, made of abstractions that spark but neither fan nor feed the flames of change. Without application, the idea of a radian is meaningless. Once applied to the issue of food deserts, however, it becomes apposite. Math offers up equally as many solutions as it does problems, which is precisely what makes math powerful: it can solve anything when applied. This cheesy, doting, poetic idea is what I think of when I hear the term "applied mathematics."

Because math's problems and solutions are two sides of the same coin, the powerful nature of math is just as commonly viewed with fear as it is with awe (a sneak peek into the paradoxicality of the next chapter). Gerrymandering is a prime example of this duality.

In the case of gerrymandering, math is both the villain and the hero, the problem and the solution. In our electioneering project in my senior year, we investigated the concepts of compactness, proportionality, and efficiency at the root of this problem. We explored the modern use of gerrymandering and its effects on today's elections. In doing this, we applied the tools of metric geometry to a reciprocal issue, giving our math knowledge purpose and power.

In the same electioneering project, we also examined the mathematics behind different voting systems around the world. We studied the electoral college, plurality, the Borda count, and more through the lens of probability. We then worked first individually and subsequently in groups to create our own voting systems, which would combine each of the positive aspects of the ones we researched. As we did so, the sense of genuine relevance, intention, and power was palpable.

With math's possible applications being so boundless, mathematicians and those who present results are responsible for using its power prudently and justly. In the case of news outlets, this is not always executed. In one brief but deeply meaningful lesson in my junior year, we learned about the skewed use of statistics in the media. Contrary to what it may seem, this usually does not refer to news outlets showing incorrect or fraudulent information but rather using accurate statistics in misrepresented or manipulated contexts to convey a particular message.

The power of math lies within its wielders and applications. Systems like project-based learning, which allow for students to identify causes meaningful to them and apply math to understand them, fully utilize and teach this incredible power. Math may be at the root of many of the world's problems, but it also provides the necessary solutions.

INVITATION TO
REFLECT

- What are some ways that you could design essential questions and find authentic audiences for the work your students are doing in your current math unit?

- In what ways can you add mathematical elements of project-based learning to your daily, weekly, and monthly practice?

CHAPTER 11

DEAR MATH,
YOU ARE PARADOXICAL

*"Dear Math, You are the reason I want to quit
with everything. Math is also the reason I feel
accomplished and proud of myself."*

WE WERE WRAPPING up a professional development session where we had put on our student hats and completed a visual pattern activity designed to help us think deeply and develop our teaching skills. An eighth grade teacher shared that she felt confident in her math skills. She had been teaching for fifteen years and had dutifully delivered content clearly and precisely to students throughout her career. The students showed up each day and took notes on identical, neatly organized binders. Every Friday, they would take a quiz. Some of the students would do well, some would not, and then the process would repeat the following week.

For our math task together, the teachers were given two clues and asked to make claims about what the clues were telling us. My co-facilitator and I intentionally designed the task to foster discourse and sense-making. We had removed some parts that made it clear and straightforward and added notches to the y-axis that ended up being 2.5 units apart—an unusual spacing but one that was necessitated from the pattern. Is this allowed? Sure!

So, this lovely woman dove into the task and kept getting caught up in the way the visual pattern connected to the graph. She was making assumptions that the notches on the y-axes were integer values and then would draw the pattern accordingly. Then we would go back to the graph, and it wouldn't connect. She worked with a colleague a bit but mostly sat with her own thoughts, seemingly growing frustrated as time went on. After we wrapped up the task, she said, "I feel like an idiot today."

I attempted to celebrate her way of diving into the problem and thinking about it in a new way, but she was too frustrated. "I don't like today," she said.

I reflected on the activity later. Had we made the task too challenging or confusing? Had we not launched the problem to ensure maximum access? Had we not asked enough supporting questions as we walked around?

I believe the bigger challenge was that this teacher was not used to having to think hard about or even be confused by problems. She was, after all, a math teacher. In many spaces, this position is revered as having utter clarity about all things math and the ability to regurgitate it when needed.

In fact, this is a gross underrepresentation of mathematics teaching. Viewing it with this straightforwardness and authority undermines math teachers' ongoing work as mathematicians themselves. It also ignores the ways that communities might come together around communal sense-making on challenging problems. This teacher was not comfortable diving into a learning pit, acknowledging that she didn't yet understand something, and then working with colleagues to find her way out of it. Instead, she created her own narrative about being an idiot and not liking the day.

As teachers, we must model the ideal mindsets for our students in our work with other teachers. We can show how to listen with intentionality to others' ideas, excitedly dive into math work that we don't yet understand, and get comfortable with the dual identities of math teacher and math learner. As we transition to sitting with this tension, we can attend more curiously to student thinking. We can use more growth language when students share ideas that don't make sense yet in the broader problem, and we can think of mathematics as a large web of interconnected ideas that we are continually exploring by heading into confusion and challenges.

PARADOX RESEARCH

A paradox is a seemingly contradictory statement that may nonetheless be true.

One famous example is called Zeno's dichotomy paradox. I often use a version of it in my calculus class when I introduce limits. I ask a student to stand up and walk halfway toward a wall in the classroom. Then, I have them walk half of the remaining distance, then halfway again. The question after they have moved a few times is, "Will this person ever reach the wall?"

The students make claims, generally in the genre of "No, because you can't keep halving and halving and end up at zero!" or "Theoretically, they won't, but in real life, they won't be able to move any further because their toes will be touching the wall." It's a beautiful visual and a way to reckon with the seemingly contradictory feeling that a person could walk toward a wall forever and never really reach it.

Another is called Simpson's paradox. In this situation, an unexpected result occurs when looking at data. Simply put, when you look at data from two variables together, you might make one claim about the trend of the data, but when you separate the variables, the opposite findings might emerge. This paradox is most frequently named in a gender discrimination case out of UC Berkeley, where the data in total showed that men were being disproportionally admitted to UC Berkeley over women. When the data was broken down by department, however, it revealed that two departments had incredibly disproportionate rates and the other departments were about even. The conclusion was that women were applying to departments that had more competitive admissions, while men more often applied to departments with fewer applicants and lower admission standards. The pooled data showed a small but statistically significant bias in favor of women.

RESPONSES TO PARADOXES

In the last couple of decades, a fair amount of research has emerged on the ways that humans handle paradoxical situations. Reportedly, people who are able to handle and even thrive in spaces where there seems to be conflict and paradoxical situations are more creative and successful. The Society for Personality and Social Psychology put out a report in 2018 stating that people who "adopt paradoxical frames, which are mental templates that encourage them to recognize and embrace contradictions" are more likely to have "creative advantages" in their thinking.

"Dear Math, In elementary school, I used to love math, but ever since math problems started getting harder and challenging me more, I didn't like it so much. I started getting distant with you around fourth grade, and I still am a little distant with you. I enjoy math when it's not too hard but not too easy (a little challenge), but ... you would just get more difficult, and I didn't like that. Now I am still distant with you, but sometimes I like solving some problems of yours." —Seventh grader

EFFECTS OF CONFRONTING PARADOXICAL SITUATIONS

The paradox mindset is not a new idea, but we can apply it to math education in new ways. Math education researcher Rochelle Gutierrez cites the need for mathematics teachers to "Play the game and change the game." How can a game be changed while you are playing it? I picture football players engaged in a game while the rules are changing in real time. It would be chaotic. However, although the rules of the game don't change very often, the strategies within each game change constantly.

While we seek to redesign math classes that center students' experiences and math identities, we can and must adapt our strategies. We must hold a paradoxical mindset where we attend to the two at the same time. We can do this work so that, as our students head to spaces that haven't yet been reformed to cultivate their mathematical brilliance, they are not left out or held back from accomplishing their dreams. This is particularly true for students who are under-privileged and need to be equipped with the language and tools to navigate the spaces. All the while, those of us with power and voices must do what we can to keep the system evolving in an equity-centered, justice-oriented way that leads to human flourishing and the use of math as a tool to make the world a better place.

Gutierrez states, "As a researcher dedicated to equity, I attempt to situate myself in 'Nepantla,' the crossroads of these tensions to highlight the phenomena at hand. Being able to name the dimensions helps us move toward highlighting tensions between the dimensions so that we might be more reflective about how we can successfully balance attending to them all." The "Nepantla" is originally a word from the Nahuatl language spoken by the Nahua people who primarily live in Central Mexico. The idea speaks to the notion of duality or holding two concepts at the same time. Gutierrez brings this idea to life, and my hope is that the word similarly gives us a way of thinking of the paradoxical feelings that mathematics gives us. It is by naming the conflicting feelings and acknowledging that those feelings can exist in tandem that we can find new, creative ways of thinking about mathematics and forge a relationship with the discipline that is decidedly whole, open, and flourishing.

Our goal is to equip students with the creative thinking fostered by understanding and accepting paradoxes as they process their feelings about math and make conjectures about the discipline of mathematics. We want students to consider statements, some of them contradictory, and then test their assumptions and provide evidence of the more justifiable ones. Sometimes, adapting assumptions can lead to new models and different (hopefully better) outcomes.

> "Dear Math, You have always been a thing in my life. You can be so fun and so frustrating at times, but whatever the case, I still need you. ... When I was younger, I used to dislike you, but as I learned and grew, I realized that you can be very fun and helpful. I've had times where you are complicated and confusing, and I've had times where you are exciting and extraordinary. There are always two sides to a leaf. My hopes for you and for me are that we can understand each other. I hope that I can do good in life, thanks to you. You are always there for me, in good and in bad. I greatly appreciate that. Math, you are one of my greatest friends, and I hope we can stay that way forever." – Calen, seventh grade

Asking students to contemplate the paradoxical sentiments that they state and to find a level of "okay-ness" in those feelings is a major step in their learning process. In terms of classroom practices, I have come to use a couple of tools throughout my career to help attend to the paradox of "loves and hates" and "can't and cans" in math. Ultimately, all these emotions are critical elements of the learning process. This is hard work, and frequently, students will look for the path of least resistance to avoid making their brains work harder.

NORMALIZING THE LEARNING PIT

UK researcher James Nottingham has done a lot of research around an idea called The Learning Pit (and wrote a book with the same name). The learning pit is an image of a young student standing at the top of one side of a pit and the "learning" is on the other side of the pit. Nottingham says that students won't learn deeply or make connections without first taking a tumble into the learning pit. It's a valuable part of the process to grow confused and then meander out of confusion. This stands in stark contrast to most students' expectations of math class: that they will try to understand something the teacher explains quickly, or they'll decide it's too hard to understand it and give up. If students come to normalize this "tumbling" process, they can hold the paradox of confusion and learning simultaneously. With my students, I have seen the power in the images and language Nottingham uses to discuss what happens as they venture into confusion. As an instructor, I have leveraged phrases like "It's okay if this isn't making complete sense to you yet; this idea is challenging and took hundreds of years for other mathematicians to explore it." Along these same lines, if we can help students celebrate when learning is hard in this way, we might have less of a shutdown when the negative stories involving "dread" and "uselessness" come up because we can more readily turn these into productive learning times.

Another way to support students in confronting the paradoxical feelings in math class is to provide frequent time for reflection with activities like graphing stories. I was first introduced to the idea of graphing stories through a Dan Meyer activity with videos of scenarios where the students would show a change in something over time. This was an integral part of my functions unit when I first started teaching, and students often enjoyed plotting these scenarios on axes. I also began leveraging graphing stories as a check-in device with my high school students and my own elementary-age children. I would ask my children at the end of the day to draw how their day was on a set of axes, most frequently labeled happiness. Image 11.1 shows my daughter with one of her graphing stories when she was in fourth grade.

Image 11.1: Poema and her graphing story about her day.

The graph provided us with a shared visual to process the day and allowed me to hear about the events and how and why they produced these feelings (the maximums and minimums on the graph). Before long, I had my students using the graphing story model to plot their feelings during the semester. They could contemplate how much they felt like they were learning, their level of happiness, the number of connections they were making, the number of ideas they were brainstorming on their warm-ups, or the number of ideas they shared on our weekly check-ins. Every activity became an opportunity for data collection and visualization in the most human-centered way because the graph wasn't the end; it was the beginning of connection and storytelling.

As students process the deep feelings they have about math class, we need to continue providing them with experiences to see the discipline in a new way and chart a path forward with this new view. As we support them in building up their self-efficacy and feelings of improvement, reflection is a critical part of the process for them and us.

> "Dear Math, Sometimes you are very fun to do. For example, geometry is one of my favorites. I really like learning with shapes. .... My past experience with you has been a mix of fun and not fun, of stressful and non-stressful. You were really easy in elementary and kindergarten, but once you go up into middle school, it gets way more difficult. You can be stressful as well if you are solving a problem that requires a lot of brain use ... if you don't know the answer and if you have tried many times but still can't get it right, ... it makes some people just give up." —Elias, seventh grade

THROUGH THE PARADOX

In your practice, you might be feeling some paradoxes. Maybe the students don't seem to understand the math content you wanted them to learn, and you feel like you want to process emotions, but you're worried about taking the time out of "instructional minutes." Maybe you want to ask students to write Dear Math letters, but you don't want to have to sit with all the troubled stories they might share. Or maybe you want to do creative and open-ended mathematics work, but you don't want your students to do poorly on a standardized test.

These apparent paradoxes are okay to sit with. Find a group of teachers and talk it through together. Share your worries about

211

the "tradeoffs." Keeping the students' experience at the center is most important as we sit with these ideas. And we must sit with them. They are the path toward improvement. We must listen to the students, design for them, and continually attend to their current needs. We are stepping in and disrupting a system that has historically not served all students well, so we need to expand our practice by discussing changes with colleagues. Change cannot happen unless we step into the feeling of the paradox and get used to holding the contradictory ideas.

"Dear Math, Math hasn't always been my favorite; in fact, I've actually struggled with it A LOT. Math has always been tricky and sometimes even frustrating to me … When I get a problem right, it makes me feel so good, and I get extremely proud of myself! Even if it doesn't happen all the time, it still makes me super happy! I don't exactly love math, but I don't hate it either. This year, I can improve my relationship with it and master it!" –Mia

WORKING THROUGH THE PARADOX IN THE CLASSROOM

It was 2020, and the COVID-19 pandemic was raging on, but the students were starting to return to school intermittently in small groups. Three teachers at my school, Juli Ruff, Amilio Aviles, and Taya Chase, were teaching on a ninth grade team and wrestling with how to support the students in processing the toll the pandemic had taken on them. They knew they couldn't act like it hadn't happened and just proceed with life as usual. They also believed that sitting around all day talking about the hard times would do nothing to lift the students out of the mire of challenges that had transformed their lives and the world. They believed in the importance of providing rigorous grade-level-appropriate content in their courses and wanted to leverage the time to do deep, meaningful work grounded in clear content goals.

Since we work at a project-based learning school, the teachers designed a project entitled "Boxing the Pandemic." They built wooden boxes that would serve as time capsules to box up memories and reflections of the time. They included a welded artifact, physical artifacts in physics, and some writing. For math, Taya encouraged the students to create a graphing story that outlined their feelings during the pandemic. They drafted what the graphs would look like and then added precision by creating piecewise functions to describe the time. The students adhered the graphing stories to the inside of the box and, on exhibition night, engaged their parents in a discussion about how the graphs helped them tell the story of the pandemic and wrestle with all the emotions of the last year. They also reflected on their work with functions and how they used Desmos to create these piecewise graphs that told their stories. For many of the students, the pandemic was paradoxical in so many ways. It was also full of pros and cons. They loved freedom, but they hated not seeing their friends. They loved the extra time with family, but they hated Zoom. They loved being able to slow down the pace of their lives, but they were struggling to find the motivation to do their work.

Being able to process their feelings in this way was beautiful and important. The students were able to name the time, add precision and narrative to the time, and not worry about trying to reconcile their various feelings. They just were. The students were holding multiple truths about the pandemic at the same time, and that was just fine.

These graphing stories and a collection of the pandemic boxes are now on display in our building as an homage to a time that changed the world and schooling forever. Students walk by and look at the boxes while shuffling between classes. The reflection and naming of the contradictions provide words and language for a time that could easily just be thought of as all one thing. I feel fortunate to work with colleagues who create spaces for students to normalize the paradoxes and support one another through the potential tension as they form a community. It is truly a beautiful thing.

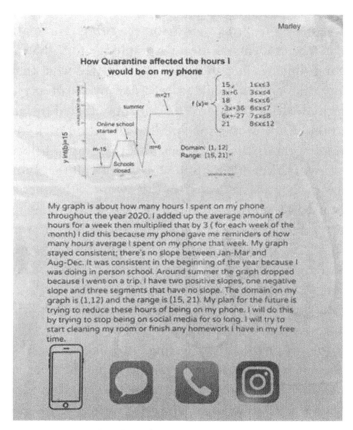

Image 11.2: Marley Shepard's quarantine graphing story reflection.

GIGI'S REFLECTION

We have spent the duration of the book first outlining the many issues with the structure of contemporary math education and, subsequently, proposing solutions. We could nicely and neatly wrap up the book just like that, with equal resolutions to the complications, a roadmap of sorts, detailing each of the mathematical pools of quicksand and whitewater rapids and which paths to take to avoid them. Instead, in one final act of mathematical theorizing, we must leave you with a fork in the road, the paradoxicality of math.

Contrary to what it may seem, I posit that the majority of this paradoxicality lies not within the particularities of philosophical math concepts (such as whether there are more decimals between zero and infinity or four and five) but, rather, within the processes of teaching and learning math. When it comes to teaching, there is a constant balancing act of pushing the student enough that they participate in genuine learning while also ensuring not to push them so much that they opt to give up instead of doing the work. It's not an easy feat. Especially when the balance correlates directly with the learning pit. Pose just the right level of challenge for a student, and they will tumble down the learning pit, brush the dirt off their shoulders, and find their way up the other side. Pose a challenge too difficult or intimidating, and they will either be too afraid to venture into the depths of the learning pit or be so blinded by the muds of confusion and discouragement in their eyes that they attempt the trek only to faceplant into the bottom, unable to think of a way up and out. In my career as a student of PBL, I have had both experiences.

Think back to Chapter 6, in which I discussed my repeated failing of a math test. This was one instance that gave me an aversion to the learning pit. I had faceplanted. In fact, I had faceplanted over and over and over. Why would I want to keep going down into the pit? I wanted to cheat my way to the knowledge on the other side, to build a bridge where I wouldn't be subject to any more mud or broken bones. Sadly, taking the bridge doesn't get you to the knowledge. It could perhaps get you to the A, but the knowledge requires the tumble. Sometimes, it even requires a faceplant or two.

When learning about systems of equations, why and how they emerge, and how to solve them, I faced the learning pit again. Although I understood the process of "substitution," I had lots of trouble with executing the "elimination" process. Because of the way Sarah structured the activities and lessons—expertly completing the balancing act—I felt brave on the edge of the pit. I tumbled, certainly, but there was no confusing or discouraging mud. Instead, I was able to use the tools around me, such as my classmates and the practice of identifying what *I know, think I know, and wonder*, to help me find my way up the other side. This resulted in my unknowingly grappling with the paradox in my mind: the learner's paradox. This dilemma involves the two opposing voices in your head. *You can do this! No, you can't.* This paradox is a perfectly normal one, and attempting to completely eradicate the negativity in your head will likely never work. Instead, we must, as learners, recognize the existence of the paradox and grow to be comfortable sitting with it. We must unpack the roots of the negativity and the steps we must take to fight it.

As a current college student, I can say that it does seem as though the learning pit deepens and widens as difficulties and pressures to perform increase. College being fairly difficult and pressurized, I am constantly faced with the learning pit, just as I'm sure you are in your everyday life. None of us can run from the learning pit. None of us can take the bridge. Everyone must work to be okay with the paradoxes within themself. We have to use the skills we've learned and practiced to climb out. We have to face the fork in the road. I believe in us. We can do it.

INVITATION TO
REFLECT

- What are some ways that you have experienced paradoxical feelings about mathematics?

- In what ways can you help students feel normalized in their seemingly paradoxical feelings about mathematics learning?

"Dear Math, We've had our ups and downs, but our story is definitely not over yet." —Taylor

CONCLUSION

RECENTLY, I GATHERED with a close community of about fifty people. We call this church, but it is a far cry from what most of us might picture when we think of a traditional church. The group included many folks who were experiencing homelessness or had experienced it in the past. Also present were middle-class people with teaching jobs, stay-at-home moms, and social justice warriors. There were immigrants from Africa, belting beautiful melodies into microphones; children throwing balls to each other; and plates being filled with carne asada. One participant frequently shouted out loud in the middle of the meal and caught everyone off guard. There was beauty and brilliance, love and laughter.

I couldn't help but think this gathering would not have existed without intentional design. Too often, our society is separated by income level, race, political affiliation, and interests, thus limiting us to the perspectives within our sphere. Our social media algorithms only exacerbate this problem by feeding us more information that is in line with our previous interests. So how do we design the places where diverse groups of people can come together and truly see and hear each other, discourse with each other, and laugh with each other, building a world full of empathy, love, and justice?

This story might seem unrelated to math class, but the same tendency to sort people into groups and classrooms with other students who are "similar in level" is pervasive in schools. It is just the tiniest microcosm of the sorting that happens in our society. About 90 percent of the time when I chat with a math teacher about their classroom, they share with me, either in asset- or deficit-based language, the wide variety of math skills or content knowledge in their classroom that makes it nearly impossible to teach every student in their class. They wonder whether a student who is struggling to make sense of fractions could and should learn alongside a student who has already mastered graphing systems of equations. If we adhere to a linear, narrow path of math learning, then we might be tempted to conclude that, no, there is no way that these students could possibly learn math in the same space. However, if we build a world where those students are valued as members of the community, are liberated to share their thinking and expertise in diverse ways, and are given mathematical dignity, they might come to see themselves as mathematicians in new and exciting ways. Having your ideas considered and wrestled with by your community helps you feel worthy and realize that you matter. All of our students are worthy of this level of dignity, and they matter wholly in whatever state they come to class.

Creating spaces through Dear Math letters and other regular check-ins with students' feelings and senses of belonging helps us to form this community. We can't design for and with students unless we start by listening to them. We must be quiet and patient. We will hear about their trauma, how they were excluded, and how they were made to feel stupid or less than. We will hear about their experiences of beauty, power, and usefulness as well. We can build on that, and we can help them heal from their trauma and find wholeness. We can design a math community where math isn't a gatekeeper but a liberator. When we ask the question, "Can we sit next to, learn from, and discourse with people who are different from us?" we help students see that, yes, we all benefit by doing this—in math class, in our school, in our community, and in life.

As I continue to ponder the notion of worldbuilding in our current education setting, I see reason for hope. I work with teachers every day who are centering student thinking, excavating student brilliance, and designing math classrooms that are communal. I also hear pushback from some who advocate for the status quo in math class, even though it didn't work for them and much of society. I am saddened by the lack of worldbuilding with these folks, and I implore them to listen to students' math stories and design with their needs at the center.

While worldbuilding can be related to constructing an imaginary world, I submit that its pursuit brings immense joy and satisfaction to us as teachers and our students as learners. Design with and for students, avoid teaching *at* them, and keep your eyes open each day for surprise moments of brilliance.

May we all find anchors to end our letters with words like:

P.S. Math, you make me feel whole.

MORE ON MATH IDENTITY FROM CHAPTER 1

MATHEMATICIAN LIKE ME

In this project, students will explore their mathematical identities as individuals and together as a classroom community. They will highlight the strengths they bring to the community based on their experiences in math classes historically and on their shared experiences. They will also learn about mathematicians throughout history and locate window-and-mirror moments in these selections. Each student's final product will be a mathematician self-portrait, including an image of their face; a collection of their strengths, areas of growth, and window-and-mirror moments; and a statement of mathematical purpose to be shared with someone in their family.

Math autobiographies

In this activity, each student tells a story about themselves as a mathematician and reflects on activities that have and have not been helpful to their mathematical learning.

Shape shifters

A student's mathematical identity plays a huge role in how they interact with math, both in school and after they leave the school environment. In this project, we will use shapes, similarity, right triangle trig, and cross-sections to explore our mathematical identities. We will explore their identity formation in their younger years, the iterative cycles they have gone through to shape their identities, their goals for the future, and the different facets of their identities that make them complex and unique. The products for this project will be shared with various audiences to help students form healthier and more positive mathematical perspectives of themselves and others.

Guardians of the math galaxy

Students will work in pairs, and each pair will create a hero who personifies the mathematical practices. The pairs participate in a competition to have their hero "look over" and "guide" the class (and the world) for the rest of the year. Then the class creates a name, superpower, costume, origin story, and archnemesis for the winning hero.

B

BELONGINGNESS BUDDIES HANDOUT FROM CHAPTER 2

Create a handout like this to give to your students so they can record the contact information and notes about their belongingness buddies.

BELONGINGNESS BUDDIES CONTACT INFORMATION

Please collect contact information for all members of your group. If you are not comfortable giving your phone number, please provide any other method of contact you are most comfortable with (for example, school email address or personal email address). Provide any particular instructions for contact under "Notes," such as "No texting, only phone calls please," or "Include images or assignments in the email," or "No texts/calls after 8 p.m."

> THIS INFORMATION SHOULD ONLY BE USED TO CONTACT OTHER STUDENTS ABOUT CLASS MATTERS. DO NOT GIVE THIS INFORMATION TO ANY PERSON OUTSIDE OF THIS CLASS. **PLEASE RESPECT YOUR FELLOW CLASSMATES AND THEIR PRIVACY!**

GROUP NAME: _____

Print First Name, Last Name	Contact Information Cell Number/ Email Address	Notes about Contact

DAILY DISCOURSE FROM CHAPTER 4

HOW TO GET STARTED WITH THIS ROUTINE

- Students arrive to class, and the teacher reviews the warm-up expectations. As the year progresses, they paraphrase and revisit them as needed. Here are some sample norms:
  - ▸ I always start by looking for something I DO know.
  - ▸ I am never done because I always ask the next question.
  - ▸ When asked, I will share my thinking with my group.
  - ▸ If asked, I will share my work with the whole class.
  - ▸ I will often make mistakes, but recognizing them is what makes my brain grow the most.

- Project the warm-up problem onto the screen with an emphasis on open-ended phrasing. Exercises described in Chapter 3 could be particularly helpful here, but other question types can work, such as:

  ‣ Say as much as you can about …

  ‣ Write as many solutions as possible to …

  ‣ How many different ways can you write …?

  ‣ What do you notice/wonder about this …?

- Give students five minutes to brainstorm and write down as much as they can about the prompt. Anything they write is a good idea, including observations, questions, connections, and examples.

- While students are working, the teacher randomly selects two students and asks them if they will be comfortable serving as the discussion leads for the day. One person will be the scribe, and the other will call on classmates who have an idea to offer.

- After five minutes of brainstorming, the teacher asks the students to circle one idea on their paper that is most noteworthy, unique, or important to them.

- For the first minute, in small groups, each student shares the idea they circled and prepares to share with the class.

- For the final four minutes, the discussion leaders guide the class in sharing out ideas and documenting them on the board. Students are encouraged to add to their notes any new information that is shared out.

- In one final closing minute, the class shares praise for the discussion leaders, highlighting specific things they did well.

- Optional assessment: on Fridays, students have a brief warm-up quiz that relies on the problems of the week and some of the ideas that were shared out during the full class discussion.

- Optional assessment: every tenth warm-up, students submit their papers with a short reflection to answer these questions:
  ‣ Which Daily Discourse were you most proud of?
  ‣ Which Daily Discourse would you want to revise?
  ‣ What did you learn from a scribe or discussion leader this week?

SEPARATING FROM GRADES

Grading has massive potential to be an area where inequity thrives. Chapter 5 covers issues around grading; however, for this routine, my main goal was to create conditions to establish a math community in the absence of grades and with the addition of regular norming, sharing, reflecting, and celebrating. The reflection forms from the students and me could be used as evidence in their portfolios, but they were in no way punitive. Including ungraded elements in the class routines was critical to creating a sense of community in my classroom.

OUTCOMES OF DAILY DISCOURSE

The outcomes of this routine are significant. From students throughout the years, I have heard that the Daily Discourse routine helps:

- put the authority in their hands
- lower anxiety around what they thought math was (pure symbols, expressions, and equations)
- create openness about discussing ideas and asking questions (this is celebrated!)
- celebrate visual thinkers
- encourage creativity, diverse ideas, and connections within abstract math problems
- empower all students to participate in and lead class discussions, reinforcing that we all have valuable contributions
- promote fluency and comfort with decontextualized and abstract math problems and mathematical concepts
- show that math is a collaborative process where we generate and build on diverse ideas and where challenges are opportunities to learn from each other
- increase confidence when sharing mathematical ideas and engaging in and facilitating mathematical discourse

SIGNS OF PROGRESS

- Weekly exit cards gauge what our students are thinking about their mathematics learning each week. One of the exit tickets I used weekly for many years had the prompt: "This week I felt comfortable sharing my ideas in class because ..." The responses to this question stayed high when I was doing Daily Discourse, which I took as a sign of success and that students were not feeling intimidated.

- Students show an overall increase in the amount of time they spend thinking about a math problem. There is a general sense that students have the tools to think through and bring their funds of knowledge to a problem. They don't rush to ask the teacher a question or wait until someone "tells them the answer." Rather, they look for things they notice, ask questions, and make connections with their peers.

- Students are doing more talking than you are. One of my colleagues once created a data collection measure called "Teacher-Talk Time." She and I worked together to record the amount of time we were talking as opposed to the collective time of the students in the class. I often say to students, "The person talking about math is learning about math. I should not be the main person talking about math in the room."

UNOPPRESSIVE ASSESSMENT FROM CHAPTER 5

SAMPLE ASSESSMENT #1 (UNIVARIATE STATISTICS)

Prompt: Give yourself 15–20 minutes to analyze this data set as thoroughly as you can. Include ideas like <u>mean</u>, <u>median</u>, and <u>mode</u>, and discuss which is the best measure for this data set. Also create a <u>box and whisker plot</u> and as many <u>histograms</u> as you have time for, and describe their shapes using words like <u>skewed</u>, <u>gaps</u>, and <u>clusters</u>. Finally (if you have time), describe how spread out the data set is with <u>range</u>, <u>standard deviation</u>, and possibly a <u>normalized bell curve</u>.

How many miles do you live from school?
0.5
1.2
1.8
1.9
2.5
4.1
5
5
6.9
7.3
7.5
10
10
10.1
11.2
12
13
14
16.1
16.7
17
17
17.2
17.6
20
45
Sum = 290.6
Count = 26
Standard Deviation = 9.1

SAMPLE ASSESSMENT #2 (RATIONAL FUNCTIONS)

Prompt: Take 10 minutes to say as much as you can about the following rational function/equation. Any annotations, graphs, tables, or questions you can add are great!

$f(x) =$ $r =$

APPENDIX

SAMPLE PORTFOLIO REQUIREMENTS FROM CHAPTER 7

FUNCTIONS PORTFOLIO

1. Summary of Learning Mind Map: Students will create a "mind map" of all the topics and information learned during this unit. It will include the idea/topic and an example or definition of that idea. The goal is to summarize and organize all your learning. Make it colorful with LOTS of visuals, arrows, diagrams, etc. If you want to keep with your theme from last time, that's great!

2. Reflection of Growth: Students will complete a reflection, citing evidence to justify each piece of growth in the reflection. All writing should be in the rubric in Google Classroom.

3. Work Samples: Students must compile all their classwork from the unit into a portfolio. Assignments should be in order and work on each should aim toward EE level work (see other side for rubric). Before submitting, have your partner check over your work and score you, and then give yourself a final score.

	Peer Assessment	Self-Assessment
1. Graphing stories day 1		
2. Graphing stories day 2		
3. Function notation and tables		
4. Uber functions		
5. Yam in the oven		
6. Desmos art masterpiece		
7. A questionable payment plan		
8. Population of the world task		
9. Exploring exponential functions with Desmos		
10. Exponential functions practice		
11. Immigrant population task		
12. Functions quizzes		

GROUPWORK ASSIGNMENTS SCORING RUBRIC

EE (exceeds expectations)	Your mathematical voice is evident. Your paper has a few ideas about ALL questions (and answers when necessary). You explain ideas thoroughly AND make connections. Asks questions. Goes beyond the above in one way or another (e.g., Desmos graph, extra practice).
ME (meets expectations)	Paper is complete; some thought process is present. Some solutions lack explanation (just numbers).
NY (not yet)	It is only partially complete (some work done). Some answers are present but not explained. No work/brainstorming is shown.
I (incomplete)	Paper is not turned in at all.

MATHEMATICAL MODELING ACTIVITIES FROM CHAPTER 9

What is a mathematical model?
Prediction:
Definition:

Define the Problem Statement:	
Define the Variables:	**Make Assumptions:**
What factors are important to consider?	*What simplifications can we make?*
What can we measure?	*How can we give ourselves a solvable problem?*
	What seems most important/ measurable?

Research and Collect Data:

What data already exists about this problem?

What data would you want to collect to better understand the problem?

How would you represent your data to help make some predictions about your problem statement?

Make an Informed Prediction:

In my research, I found …

Based on my research, I predict …

Compare Results to Prediction:

Collect your own data and see how accurate your prediction was.

Analysis of the Model:

What are the strengths and weaknesses of your model?

What will you do next to make your model even better?

What changes would you make to the variables or data?

Amended from *Math Modeling: Getting Started and Getting Solutions* published by SIAM

ABOUT THE AUTHORS

Sarah Strong

Sarah Strong loves hearing people's math stories. She has taught math to grades six through twelve at High Tech High in San Diego, and she also works for the High Tech High Graduate School of Education, teaching Math Methods and Advanced Math Pedagogy courses and supporting the new math teachers in the organization. She has led workshops on project-based learning in mathematics, on student-centered assessment, and on alternative grading systems. After designing and facilitating a project on math identity in 2017, Sarah grew interested in the ways students told stories about their experiences in math class. Ever since, she has been accumulating these beautiful stories and using them to design classroom experiences that center students wholly. She loves camping with her husband and their ten- and twelve-year-old children, and she is a runner and a bread baker.

Gigi Butterfield

Gigi Butterfield is currently a screenwriting major at Loyola Marymount University and a graduate of Gary and Jerri-Ann Jacobs High Tech High in San Diego. She is recovering from her fraudulent fondness of mathematics and thrives in situations where she can explore math deeply and ask thoughtful questions of her peers and teachers. Gigi has attended project-based learning schools since the age of five and, even in college, is passionate about how PBL plays an integral role in revitalizing heavily antiquated math pedagogies. She was captain of the basketball team, head of student ambassadors, leader of Model United Nations, and a member of Student Senate. She is a Jeopardy fanatic and is hoping to go into comedy writing. No better start to a comedy career than with a dissertation on the reimagining of math education!

ACKNOWLEDGMENTS

From Sarah:

This book would not exist without the prompting of my number one teammate in life, Mike Strong. From your initial "You two should write a book" in my classroom over lunch with Gigi, to our Daily Discourse about the future of education as a means to justice and the end of poverty, you are my hero and my inspiration. Thank you for providing me with unwavering love and devotion personally and with cognitive dissonance and criticality professionally.

To my children, Poema and Canon, you are my joy and push me to continue learning each day. I love you with my whole life.

To my parents, Dana and Karrie Richardson, thank you for creating a home that encouraged me to act justly, love mercifully, and walk humbly. To my mom, thanks for sharing your math story with me for this book and for being open to continuous learning. And to my dad, thanks for continuing to pour your life into young people and their learning even after you retired.

To the rest of my family, Jesse and Leanne, Sarah and Kevin, Diana, Harlowe, and Marenn and David, I love you.

To my mentors, Cate Challen, Katerina Milvidskaia, Chris Nho, Bryan Meyer, Brian Delgado, Curtis Taylor, Daisy Sharrock, Ryan Gallagher, Kristin Komatsubara, Juli Ruff, Melissa Daniels, Kaleb Rashad, and Sarah Fine, you have continually pushed me and formed me into the teacher I am today. Thank you for believing in me.

To Dylan Bier, thank you for helping me design the shape shifters project that birthed the Dear Math letters and started my journey into exploring mathematical identity.

To all my teaching partners, thanks for pushing me to design projects that provide the best possible learning experience for students each day.

To my many pre-service teacher students, thanks for keeping me honest and teaching me new things each day.

To the math team at HTH, thank you for carrying the fire, for continually believing in the mathematical brilliance of all our students, and for believing that math doesn't have to be a gatekeeper but can be explored in diverse communities of learners. Keep worldbuilding.

To Alisa Ward, Claire Wolf, Michelle Nho, Miok, Serena Talbadillo, Sarah Looysen, Sarah Morrison, Shelby Robinson, Jessica Walton, Alli Greenaway, and my BFF Alexis Gwin Miller, thank you for being my thought partners in life.

To my Ebenezer Family, thank you for always pushing me to learn and more deeply understand the need to end poverty and injustice.

Finally, to everyone at Times 10 Publications, thank you for believing in this book, in its message, and that we must truly center student voices and student brilliance in education work. May there be many more works like this.

From Gigi:

To my parents, Amy and Robert, thank you for loving, supporting, and pushing me. The fondness for learning and drive for growth that you have instilled within me are the reason I view the world with intrigue and possibility, and the reason this book exists.

To my sisters, Riley and Quinn, thanks for helping to shape the childhood and home that I reflect on with genuine appreciation.

To all my former teachers at Explorer Elementary, High Tech Middle Media Arts, and High Tech High, thank you for furthering and celebrating my love of learning.

To my accomplices in life, Noah Borkum, Parker Frost, Kieran Saucedo, Chloe Slack, Taylor Paris, Daniel Solot, and Reilly Talbot, thanks for surrounding me with laughter, joviality, and meaningful discourses.

And to Times 10 Publications, thank you for recognizing the commonly unsung importance of math in many facets of life and for believing in Sarah's and my ability to help sing it louder.

REFERENCES

Chapter 1

Adichie, Chimamanda. 2009. "The Danger of a Single Story." July 2009. TED Talk. https://www.ted.com/talks/chimamanda_ngozi_ adichie_the_danger_of_a_single_story?language=en.

Ashcraft, MH, Moore, AM. 2009. "Mathematics Anxiety and the Affective Drop in Performance." *Journal of Psychoeducational Assessment* 27, 3:197–205. https://doi. org/10.1177/0734282908330580.

Betz, N.E. 1978. "Prevalence, distribution, and correlates of math anxiety in college students." *Journal of Counseling Psychology* 25, 5, 441–448. https://doi.org/10.1037/0022-0167.25.5.441.

Boaler, Jo. 2017. "It's time to stop the clock on Math Anxiety. Here's the latest research on how." *The Hechinger Report*, April 3, 2017. https://hechingerreport.org/ opinion-time-stop-clock-math-anxiety-heres-latest-research/.

Gordon, Sheldon P. 2008. "What's Wrong with College Algebra?" *PRIMUS 18*, issue 6:516–541. https://doi. org/10.1080/10511970701598752.

Wenger, E. 1998. *Communities of practice: Learning, meaning and identity.* Cambridge, UK: Cambridge University Press.

Chapter 2

Beechum, N.O. 2012. "Teaching adolescents to become learners. The role of noncognitive factors in shaping school performance: A critical literature review." Chicago: University of Chicago Consortium on Chicago School Research. https://consortium. uchicago.edu/publications/teaching-adolescents-become-learners-role-noncognitive-factors-shaping-school.

Farrington, C.A., Roderick, M., Allensworth, E., Nagaocka, J., Keyes, T.S., Johnson, D.W., & Lambert, Rachel. 2019. "The Myth of Low Kids and High Kids." November 15, 2019. CMC-South Ignite Event in Palm Springs. https://www.youtube.com/watch?v=8nQHDin3pmQ.

Parks, Amy Noelle. 2010. "Metaphors of hierarchy in mathematics education discourse: the narrow path." *Journal of Curriculum Studies*: 42, 1:79–97, First published on Sept. 7, 2009 (iFirst). http://dx.doi.org/10.1080/00220270903167743.

Chapter 3

Gutstein, Eric. 2006. *Seeing the World with Mathematics*. New York: Routledge.

Harel, Guershon. 2007. "The DNR system as a conceptual framework for curriculum development and instruction." In *Foundations for the future of mathematics*. New York: Routledge. https://www.math.ucsd.edu/~harel/DNRleshBook.pdf.

Meyer, Dan. 2015. "If Math Is The Aspirin, Then How Do You Create The Headache?" *dy/dan Blog*. June 16, 2015. https://blog.mrmeyer.com/2015/if-math-is-the-aspirin-then-how-do-you-create-the-headache/.

Chapter 4

Boaler, Jo. 2019. *Limitless Mind*. New York: HarperOne.

Whyte, Julie, Anthony, Glenda. 2012. "Maths anxiety: The fear factor in the mathematics classroom." *New Zealand Journal of Teachers' Work*. Vol 9, Issue 1, 6–15. https://www.researchgate.net/publication/260710496_Maths_ anxiety_The_fear_factor_in_the_mathematics_classroom.

Chapter 5

Anderson, Melinda D. 2017. "How Does Race Affect a Student's Math Education?" *The Atlantic*. https://www.theatlantic.com/ education/archive/2017/04/racist-math-education/524199/.

Battey, Dan, Leyva, Luis. 2016. "A Framework for Understanding Whiteness in Mathematics Education." *Journal of Urban Mathematics Education*. Vol 9, No. 2, 49. https://doi.org/10.21423/jume-v9i2a294.

Freire, Paulo, 1921–1997. *Pedagogy of the Oppressed*. New York: Continuum, 2000.

Martin, Danny. 2018. "Taking a knee in math education." May 2018. NCTM annual conference in Washington DC.

Principles to Actions: Ensuring Mathematical Success for All. 2014. Reston, VA:NCTM, National Council of Teachers of Mathematics.

Chapter 6

Ball, Deborah. 2018. "How can mathematics teaching disrupt racism and oppression?" June 2, 2018. Austin, TX: National Inquiry-Based Learning Conference. https://static1.squarespace. com/static/577fc4e2440243084a67dc49/t/5b155ab26d2a734db 2ced792/1528126134248/060218_NIBL.pdf.

Boaler, Jo. 2019. *Limitless Mind*. New York: HarperOne.

Common Core State Standards Mathematics. 2010. Washington, DC: National Governors Association Center for Best Practices, Council of Chief State School Officers.

Horn, Ilana Seidel. 2012. *Strength in Numbers: Collaborative Learning in Secondary Mathematics.* Reston, VA: National Council of Teachers of Mathematics.

Chapter 7

Berger, Ron. *An Ethic of Excellence: Building a Culture of Craftsmanship with Students.* Portsmouth, NH: Heinemann, 2003.

Su, Francis. 2020. *Mathematics for Human Flourishing.* New York: Yale University Press.

Takahashi, Akihiko. 2021. *Teaching Mathematics through Problem Solving: A pedagogical approach from Japan.* New York: Routledge.

Chapter 8

Oh, Travis Tae. 2021. "What Is the Underlying Psychology of Having Fun?" Published in *Psychology Today* on June 29, 2021. https://www.psychologytoday.com/us/blog/the-pursuit-fun/202106/what-is-the-underlying-psychology-having-fun.

Robinson, Lawrence, Smith, Melinda, Segal, Jeanne, Shubin, Jennifer. 2021. "The benefits of play for adults." Published on Help Guide, July 2021. https://www.helpguide.org/articles/mental-health/benefits-of-play-for-adults.htm.

Chapter 9

Boaler, J., Cordero, M., & Dieckmann, J. 2019. *Pursuing Gender Equity in Mathematics Competitions. A Case of Mathematical Freedom.* MAA, *FOCUS*, Feb/March 2019.

Chapter 10

Berry, Robert Quinlyn, III, Conway, Basil M, IV, Lawler, Brian, Staley, John W. 2020. *High School Mathematics Lessons to Explore, Understand, and Respond to Social Injustice.* Thousand Oaks, CA: Corwin.

Petty, Nicola. 2018. "Mathematics is powerful." Published on Creative Maths in 2018. https://creativemaths.net/blog/mathematics-is-powerful/.

Chapter 11

Boaler, J., Cordero, M., & Dieckmann, J. 2019. *Pursuing Gender Equity in Mathematics Competitions. A Case of Mathematical Freedom.* MAA, *FOCUS*, Feb/March 2019.

Gutierrez, Rochelle. 2009. *Play the Game, Change the Game.* Teaching for excellence and equity in mathematics. TODOS, Vol 1, No. 1, Fall 2009.

Leung, Angela. 2018. *Why confronting paradoxes can give you a creative boost.* Society for Personality and Social Psychology: Character and Context. July 10, 2018.

Nottingham, J.A. 2007. *Exploring the Learning Pit. Teaching Thinking and Creativity*, 8:2(23), 64–68. Birmingham, UK: Imaginative Minds.

Schad, Jonathan, Miron-Spektor, Ella, Lewis, Marianne W. 2020. Paradoxes of Change and Changing through Paradox. *Palgrave Handbook of Organizational Change Thinkers (2nd edition).* Palgrave Macmillan.

MORE FROM
TIMES 10 PUBLICATIONS

Browse all titles at 10Publications.com

Hacking School Discipline
9 Ways to Create a Culture of Empathy & Responsibility Using Restorative Justice
By Nathan Maynard and Brad Weinstein

Reviewers proclaim this *Washington Post* Bestseller to be "maybe the most important book a teacher can read, a must for all educators, fabulous, a game changer!" Teachers and presenters Nathan Maynard and Brad Weinstein demonstrate how to eliminate punishment and build a culture of responsible students and independent learners in a book that will become your new blueprint for school discipline. Eighteen straight months at #1 on Amazon and still going strong, *Hacking School Discipline* is disrupting education like nothing we've seen in decades—maybe centuries.

Hacking Student Learning Habits
9 Ways to Foster Resilient Learners and Assess the Process, Not the Outcome
By Elizabeth Jorgensen

Traditional outcome-based grades make school a place of right or wrong answers—a rigid system that impedes enjoyment and learning. In contrast, innovative teachers of all subjects and grade levels use process-based assessment to build positive classroom cultures and help students focus on the learning, not the grades. Award-winning writer and teacher Elizabeth Jorgensen shows how to create process-based assessments that help students develop habits of higher-order thinking. It is about embracing, trying, failing, and trying again. It is about turning "What did you get on the test?" into "How did you get that on the test?"

Browse all titles at 10Publications.com

Hacking Project Based Learning

10 Easy Steps to PBL and Inquiry in the Classroom

By Ross Cooper and Erin Murphy

As questions and mysteries around PBL and inquiry continue to swirl, experienced classroom teachers and school administrators Ross Cooper and Erin Murphy empower those intimidated by PBL to cry, "I can do this!" while providing added value for those who are already familiar with the process. *Hacking Project Based Learning* demystifies what PBL is all about with ten Hacks that construct a simple path that educators and students can easily follow to achieve success.

Project Based Learning

Real Questions. Real Answers. How to Unpack PBL and Inquiry

By Ross Cooper and Erin Murphy

Educators would love to leverage project based learning to create learner-centered opportunities for their students, but why isn't PBL the norm? Because teachers have questions. *Project Based Learning* is Ross Cooper and Erin Murphy's response to the most common and complex questions educators ask about PBL and inquiry, including: How do I structure a PBL experience? How do I get grades? How do I include direct instruction? What happens when kids don't work well together? Learn how to teach with PBL and inquiry in any subject or grade.

Browse all titles at 10Publications.com

Hacking Mathematics
10 Problems that Need Solving
By Denis Sheeran

What if every one of your students loved math? Now you can make the impossible a reality, and your students will race to complete your math problems. In *Hacking Mathematics*, teacher, author, and math consultant Denis Sheeran shows you how to hack your instructional approach and assessment procedures to promote an amazing culture of mathematical inquiry and engagement that few students ever see.

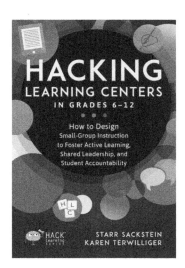

Hacking Learning Centers in Grades 6–12
How to Design Small-Group Instruction to Foster Active Learning, Shared Leadership, and Student Accountability
By Starr Sackstein and Karen Terwilliger

Learning centers are dynamic spaces where students can become robust thinkers, problem-solvers, and brave leaders. *Hacking Learning Centers in Grades 6–12* shares *why* and *how* to design small-group instruction that includes everyone and encourages students to collaborate, experiment, reflect, self-assess, and transfer the learning to their lives beyond school. Starr Sackstein and Karen Terwilliger show how learning centers empower unexpected leaders, raise the bar on student accountability, activate the fun to bring learning to life, and inspire students to share ideas and make decisions.

Browse all titles at 10Publications.com

RESOURCES
FROM TIMES 10 PUBLICATIONS

Nurture your inner educator:
10publications.com/educatortype

Podcasts:
hacklearningpodcast.com
jamesalansturtevant.com/podcast

On Twitter:
@10Publications
@HackMyLearning
#Times10News
#RealPBL
@LeadForward2
#LeadForward
#HackLearning
#HackingLeadership
#MakeWriting
#HackingQs
#HackingSchoolDiscipline
#LeadWithGrace
#QuietKidsCount
#ModernMentor
#AnxiousBook
#HackYourLibrary

All things Times 10:
10publications.com

x10

TIMES 10 PUBLICATIONS provides practical solutions that busy educators can read today and use tomorrow. We bring you content from experienced teachers and leaders, and we share it through books, podcasts, webinars, articles, events, and ongoing conversations on social media. Our books and materials help turn practice into action. Stay in touch with us at 10Publications.com and follow our updates on Twitter @10Publications and #Times10News.

Made in the USA
Monee, IL
15 September 2022

14033372R00149